Pushing Electrons

A Guide for
Students of Organic Chemistry

Second Edition, Updated Version

Pushing Electrons

A Guide for
Students of Organic Chemistry

Second Edition, Updated Version
Formerly *Electron Movement: A Guide for Students of Organic Chemistry*

Daniel P. Weeks, Ph.D.

Northwestern University

Saunders College Publishing
A Harcourt Brace College Publisher
Fort Worth   Philadelphia   San Diego   New York   Orlando   Austin
San Antonio   Toronto   Montreal   London   Sydney   Tokyo

Text Typeface: Times Roman
Compositor: TCSystems, Inc.
Acquisitions Editor: John Vondeling
Developmental Editor: Sandra Kiselica
Managing Editor: Carol Field
Project Editor: Laura Shur
Manager of Art and Design: Carol Bleistine
Art Director: Robin Milicevic
Text Designer: Anne O'Donnell
Cover Designer: Lawrence R. Didona
Director of EDP: Tim Frelick
Production Manager: Carol Florence
Marketing Manager: Marjorie Waldron

Cover Credit: Artwork by Lawrence R. Didona

Printed in the United States of America

PUSHING ELECTRONS: A GUIDE FOR STUDENTS OF ORGANIC CHEMISTRY,
Second Edition, Updated Version

0-03-011652-X

Library of Congress Catalog Card Number: 91-050634

567890123   016   1098765432

This book was printed on paper made from waste paper, containing 10% post-consumer waste and 40% pre-consumer waste, measured as a percentage of total fiber weight content.

To my sons,
John and Timothy

Postcard from a Student

Dear Dan,
Last night I was sitting on the shore of San Francisco Bay watching the fog roll in and suddenly I understood what organic chemistry was all about.

Ric

Preface

Among the factors that set organic chemistry apart from the other traditional divisions of chemistry are the endless variety of structure, the strong dependence on the resonance theory, and the emphasis on mechanisms. As students become more facile with structure and more adroit at writing resonance structures and mechanisms, organic chemistry changes from a bewildering array of facts to a unified science. Newly encountered reactions can be understood from previously learned principles.

At the introduction of resonance or mechanisms, the professor begins using little arrows that resemble fishhooks ($\curvearrowright$, $\curvearrowright$). Using these arrows, electrons are moved around molecules and appear—as if by wizardry—precisely where the professor needs them. The penchant for moving electrons has earned organic chemists the pejorative *electron pusher* that, like *Yankee Doodle,* we wear proudly.

I have observed that students avoid pushing electrons. They find it difficult, therefore, to write resonance structures and lose a powerful tool to explain reactivity. Many students never become comfortable with bond-making and bond-breaking steps in organic mechanisms. With this low-tech, pencil-and-paper program a student can learn to push electrons with the same wizardry as the professor.

While working through this book, the student will learn to construct Lewis structures of organic molecules and functional groups. The functional groups will be learned early and in a way that minimizes the necessity for memorization. Then a systematic method is presented whereby all the possible resonance structures for a molecule, ion, or radical can be generated by beginning with one Lewis structure and moving electrons. Finally, patterns of bond making and bond breaking are presented to show how organic mechanisms can be constructed. The programmed approach emphasizes repetition and active participation in learning how to push electrons.

In writing the second edition, I received helpful comments from Dr. Joyce C. Brockwell and Professor Emeritus Charles D. Hurd (Northwestern University), Dr. Kenneth Turnbull (Wright State University), and Dr. Darrell J. Woodman (University of Washington). The manuscript was typed by Ms. Sherri Smith who is just about to become a teacher. She will improve the state of American education. The first edition would not have been written except for Leslie Schweitzer. I thank Sandi Kiselica and Martha Brown of Saunders College Publishing for their patience and support during the editorial process.

Daniel P. Weeks
November 1994

To the Professor

Reviewing a manuscript that you wrote 17 years ago is, at once, humbling and comforting. In this case, I have been more comforted than humbled. In rewriting my *parvum opus* I conclude that I got it mostly right the first time.

Students will be better at organic chemistry if they master the skills in *Pushing Electrons*. Originally, the book was based on a number of sound educational principles, some of which I didn't understand at the time. I chose to limit the book to teaching one, crucial skill. I chose to keep it short, methodical, and repetitious.

You will appreciate how tempting it was to add to the book, especially when reviewers encouraged it. But *Pushing Electrons* is a supplement. It knows its place. A student will complete it in 10 to 14 hours and have the rest of the time to do important things such as listening to you and studying the main text.

The approach is methodical, the pace is slow. Strong students will race through certain sections, but they will benefit. I tell them in the text, "you can never be too skilled at pushing electrons." Weaker students will appreciate that the author still holds a grudge against his calculus professors who skipped steps.

Just as repetition teaches how to hit a baseball, it teaches how to write Lewis structures and how to push electrons properly. An unintentional benefit accrues as students become comfortable writing and rewriting organic structures. The book won't eliminate pentavalent carbon, but it will reduce the frequency of its appearance. The structures in this book, except for rings, show all carbons and hydrogens. I assert that the early introduction of line structures does students great harm.

The book has its own organization; it is not tied to that of any text. Assign Chapter 1, Lewis Structures, right away. Chapters 2 and 3 are relatively independent of one another; assign them as you see fit. It is not necessary for students to complete Chapter 2 before beginning Chapter 3. Although a chapter may be interrupted, it should be done from beginning to end.

Finally, in the second edition I have added sections on free radicals because of their resurgence in organic synthesis. Users of the first edition will note that unshared electron pairs are now shown as dots instead of those embarrassing dashes. I was taught using dashes, and they were so much easier to type before the days of word processing.

To the Student

I have only three instructions. First, supply an answer wherever a line appears under a blank space. The correct answer might be a word, a number, a structure, or some arrows. Second, don't look up the correct answers until you have made a serious try at doing it yourself. Third, if you plan to just look up the right answers and transcribe them, return the book and get your money back.

This program uses two effective learning devices: active involvement and repetition. You will participate actively in the learning process. Because so much of the academic experience consists of receiving information, it should be refreshing to work through a program using your own wits. You will see an example of an operation and then carry it out several times as the supporting material is gradually removed. The approach is methodical. Some of you will find that you can accelerate your trip through certain sections. But the program has been written in the hope that none of you will ever feel abandoned. Expect to spend a total of 10 to 14 hours to complete the program.

I considered not providing the answers to the questions, but they really are essential to the proper use of the book. Only after you have worked through a section should you check the answers. Check carefully for, as in all science, precision is important.

Contents

Glossary

Boldfaced numbers refer to the page on which the term appears.

Aliphatic   One of two broad classes of organic compounds. The other is *aromatic*. Aliphatic compounds are composed of chains of carbon atoms. A subgroup within this class is the alicyclic compounds that contain rings of carbon atoms but are similar to open-chain compounds in chemical and physical properties. Actually, the most sensible definition of aliphatic compounds is "compounds that are not aromatic."   **16**

Aromatic   One of two broad classes of organic compounds. The other is *aliphatic*. Benzene, compounds containing a benzene ring, and compounds similar to benzene in chemical properties constitute the aromatic compounds.   **16**

Carbanion   An ion in which a negative charge resides on a carbon atom.   **34**

Carbocation   An ion in which a positive charge resides on a carbon atom.   **34**

Double bond   The combination of a sigma bond and a pi bond between two atoms. The carbon–carbon double bond is written $C = C$. In this notation there is no convention regarding which line represents the sigma and which, the pi bond. The presence of a double bond makes a compound unsaturated.   **7**

Electron attracting and electron releasing groups   Various groups of atoms found in organic compounds are electron attracting or releasing, relative to the carbon atom to which they are attached. The presence of these groups can have a marked effect on chemical reactivity. Some electron releasing groups are: $-NH_2$, $-OCH_3$, and $-CH_3$. Some electron attracting groups are: $-NO_2$, $-COOH$, $-F$, $-C1$, $-Br$, and $-CN$.   **74**

Electrophile   The word comes from the Greek meaning "electron lover." Electrophiles are chemical entities (atoms, ions, radicals, or molecules) that seek electrons. Electrophiles are electron-poor. They are designated as E or E+. When an electrophile reacts with some electron-rich center on an organic molecule, the process is called electrophilic attack.   **72**

Formal charge   A Lewis structure is, after all, only a graphic representation of a molecular structure. As useful as they are, Lewis structures are not entirely descriptive of a molecule. The formal charge is the charge that would be on an atom in a molecule if the Lewis structure were an entirely accurate representation of the molecule. In one sense, formal charge is simply electron bookkeeping. However, the location of formal charge on a Lewis structure is always revealing of the chemistry of the molecule.   **20**

Functional group   One can think of organic compounds as being composed of two parts: a hydrocarbon backbone and a functional group. The latter is an atom or group of atoms that confer some distinct chemical or physical property. The functional group is the site of chemical reactions. Thus, molecules with the same functional group undergo similar chemical reactions despite having quite different hydrocarbon backbones.   **22**

Grignard reagent   A reagent formed by reaction of an alkyl halide with magnesium in an either solvent. These reagents are extremely useful and versatile intermediates for synthesizing organic compounds. The actual structure of the Grignard reagent is rather complex. It is expedient, however, to consider it an ion pair, as this text does.   **97**

Heterolytic cleavage (heterolysis)   The breaking of a bond so that the electrons that formed the bond are distributed unevenly between the two fragments. Heterolysis results in an anion and a cation.   **27**

$$A\overset{\curvearrowright}{-}B \longrightarrow A^+ + :B^-$$

or

$$A\overset{\curvearrowleft}{-}B \longrightarrow A:^- + B^+$$

The counterpart of heterolysis is *homolysis,* in which the electrons are distributed evenly between the two fragments. The result is a pair of radicals.

$$A\overset{\curvearrowright\curvearrowleft}{-}B \longrightarrow A\cdot + \cdot B$$

Nucleophile   The word comes from the Greek, meaning "nucleus lover." Nucleophiles are chemical entities (ions or molecules) that seek an electron-poor center. Nucleophiles are electron-rich. They are designated as Nu or Nu⁻. When a nucleophile reacts with some electron-poor center of an organic molecule, the process is called *nucleophilic attack*.   **72**

Pi bond and pi electrons Bonds that are the result of lateral (side to side) overlap of atomic p-orbitals. Pi bonds are weaker than sigma bonds. Pi bonds are always found in combination with a sigma bond. Therefore, single bonds are never pi bonds. Pi electrons are the electrons in pi bonds. **50**

Carbocation Carbocations are classified as primary, secondary, or tertiary according to the following scheme.

Primary The carbon atom with the positive charge is *directly* attached to only one other carbon atom and two hydrogen atoms. For example,

$$CH_3-\overset{+}{C}H_2$$

Secondary The carbon atom with the positive charge is directly attached to two other carbon atoms and one hydrogen atom. For example,

$$CH_3-\overset{+}{C}H-CH_3$$

Tertiary The carbon atom with the positive charge is directly attached to three other carbon atoms and no hydrogen atoms. For example,

$$\overset{\displaystyle CH_3}{\underset{\displaystyle CH_3}{CH_3-\overset{|}{\underset{|}{C}}{}^+}}$$

The order of stability of carbocations is tertiary > secondary > primary. **123, 124**

Protonation The combination of a proton (hydrogen ion) with some organic base to give the conjugate acid of that base. **90**

$$H^+ \; + \; :B \; \longrightarrow \; H-B^+$$

Saturated Compounds that contain only single bonds cannot add other reagents. Thus, they are called *saturated*. **1**

Secondary carbocation (See Carbocation.)

Sigma bond and sigma electrons Bonds that are the result of axial overlap of atomic orbitals. All single bonds are sigma bonds. Sigma electrons are the electrons in sigma bonds. **88, 96**

S_N1 and S_N2 reactions Organic reactions can be classified according to their mechanisms. The S_N1 and S_N2 reactions are probably the best known of these. This classification gives: the result, the type of reagent, and the molecularity of the reaction. Thus, S_N1 stands for "substitution, nucleophilic, unimolecular," and S_N2 stands for "substitution, nucleophilic, bimolecular." **88, 95**

Stereochemistry   The study of the arrangement of atoms in three-dimensional space and how the arrangement affects the chemical and physical properties of molecules.   **95**

Steric strain   Strain (instability) present in a molecule because of the arrangement of the atoms. Steric strain usually arises because bond angles are forced to depart from their ideal value, or because nonbonded atoms are required to be too close to one another.   **49**

Tertiary carbocation   (See Carbocation.)

Unsaturated   Compounds that contain double or triple bonds are capable of adding other reagents. Thus, they are called *unsaturated.*   **6**

Unshared electrons   Valence electrons in a molecule that are not involved in binding one atom to another. In a Lewis structure they are the exclusive property of one atom. They are also called *nonbonding* electrons.   **4**

Lewis Structures

1

In a perfect world there would be no need for this chapter at the beginning of a course in organic chemistry. The skills described here are taught in general chemistry. Skip this chapter if you feel entirely confident that you are able to write Lewis structures correctly and are comfortable with the concepts of ions, radicals, multiple bonding, and formal charge. On the other hand, these skills are vital to organic chemists and a little review never hurt anyone.

The distinguished American chemist, Gilbert Newton Lewis (1875–1946), was the first to propose that atoms share a pair of electrons to form a covalent bond. Lewis electron-dot formulas are familiar to all students of introductory chemistry. Lewis used dots to denote all valence electrons in chemical formulas. We will use the current convention that denotes a pair of shared electrons (a covalent bond) as a dash connecting the atoms and shows only the unshared electrons as dots.

Using Lewis's notation together with a judicious application of the octet rule allows us to write useful chemical formulas. The most important requirement for the formation of a stable compound is that the atoms achieve a closed shell, noble gas configuration. For all but the very light elements, this means eight electrons. The light elements (H, Li, Be) require two electrons, the configuration of helium. A few compounds such as borane (BH_3), boron triflouride (BF_3), and aluminum chloride ($AlCl_3$), which are important to elementary organic chemistry, do not conform to the octet rule.

Molecules and Functional Groups

Saturated Groups

Proper Lewis structures for **saturated** compounds can be written by following this brief set of rules.

RULE 1

Write the molecular skeleton. Historically, the correct skeleton of each functional group was deduced from experiment. Most of that has been done, however, and we shall assume that you know, or will know shortly, that methane's skeleton is

H

H   C   H   and not   H   H   C   H   H

H

and formic acid's molecular skeleton is

O

H   C   O   H   and not   H   C   O   O   H

which is often written to save space. In your initial study of organic chemistry you must be careful to not let space-saving notation mislead you about structure.

RULE 2

Assume that all bonds are covalent. This assumption is not always accurate, but it works most of the time, especially for organic compounds. As your experience increases, you will learn when to use ionic bonds.

RULE 3

Count the available valence electrons. Each atom in a compound brings into the union a certain number of valence electrons. The number is determined by the atom's Group in the periodic table.

1. Hydrogen is a Group __I__ element and each hydrogen will contribute __1__ valence electron. Carbon is a Group _____ (Roman numeral) element and each carbon will contribute _____ (number) electrons. Every oxygen atom in a compound will contribute _____ valence electrons.

    Chloromethane has the molecular formula CH_3Cl. Its skeleton is

    H

    H   C   Cl

    H

    and the number of valence electrons may be determined as follows. There are three hydrogen atoms, each of which contributes __1__ valence electron; the single carbon contributes

___4___ electrons; and the single chlorine atom contributes ___7___ electrons, making a total of
___14___ . A convenient tabular form of this calculation is

$$
\begin{array}{lll}
3 & H & 1 \times 3 = 3 \\
1 & C & 4 \times 1 = 4 \\
1 & Cl & 7 \times 1 = \underline{7} \\
& & \qquad\quad 14
\end{array}
$$

2. Methanol has the molecular formula CH_4O. Its skeleton is

$$
\begin{array}{c}
H \\
H \quad C \quad O \quad H \\
H
\end{array}
$$

Each of four hydrogen atoms contributes _____ valence electron, the carbon atom contributes_____,
and the oxygen atom contributes _____, making a total of _____.

$$
\begin{array}{llll}
\underline{\qquad} & H & \underline{\qquad\qquad} & = \\
\underline{\qquad} & C & \underline{\qquad\qquad} & = \\
\underline{\qquad} & O & \underline{\qquad\qquad} & = \underline{\qquad}
\end{array}
$$

RULE 4

*Add electrons to the skeleton by making single bonds between the atoms that are bonded and then by
providing each atom with a complement of eight electrons* (hydrogen requires only two electrons).

3. The skeleton of chloromethane is

The central carbon atom is bonded to each of the other atoms by a shared electron pair
(represented by a straight line, —) giving

$$
\begin{array}{c}
H \\
| \\
H - C - Cl \\
| \\
H
\end{array}
$$

Now, each hydrogen has two electrons and the carbon atom has eight. However, chlorine must be provided with **unshared electrons** (represented by pairs of dots, ••) to complete its octet, thus,

$$
\begin{array}{c}
\text{H} \\
| \\
\text{H}-\text{C}-\overset{\displaystyle ..}{\underset{\displaystyle ..}{\text{Cl}}}\text{:} \\
| \\
\text{H}
\end{array}
$$

4. Methanol's skeleton is

————————

Connecting all bonded atoms by means of an electron pair (single bond) gives

————————

and completing the octet of oxygen with two pairs of dots gives

————————

RULE 5

Count the electrons in the Lewis structure and compare the result with the number derived from Rule 3. If the two numbers are the same, the Lewis structure is correct.

5. The structure for chloromethane is

$$
\begin{array}{c}
\text{H} \\
| \\
\text{H}-\text{C}-\overset{\displaystyle ..}{\underset{\displaystyle ..}{\text{Cl}}}\text{:} \\
| \\
\text{H}
\end{array}
$$

It contains __14__ valence electrons. The number of available valence electrons in chloromethane is _____. The Lewis structure is <u>correct</u>.

6. The structure for methanol is

It contains _____ valence electrons. The number of available valence electrons is _____. The structure is _____ (correct, incorrect).

Exercises

Using the method outlined above, derive the Lewis structures for the following compounds. The unbonded skeletons are provided.

7. Dimethyl ether (C_2H_6O)

Rule 3   **Rule 4**

_____ H _____ =   H   H

_____ C _____ =   H C O C H

_____ O _____ = _____   H   H

———————

Rule 5

No. of electrons in structure _____
No. of valence electrons _____
Structure is _____ (correct, incorrect)

8. Methylamine (CH_5N)

Rule 3   **Rule 4**

_____ H _____ =   H   H

_____ C _____ =   H C N

_____ N _____ = _____   H   H

———————

Rule 5

No. of electrons in structure _____

No. of valence electrons _____

Structure is _____

9. Methanethiol (CH_4S)

              Rule 3   **Rule 4**

_____ H _____ =   H

_____ C _____ =   H C S H

_____ S _____ = _____   H

                                    =====

Rule 5

No. of electrons in structure _____

No. of valence electrons _____

Structure is _____

10. Methylal ($C_3H_8O_2$)

              Rule 3   **Rule 4**

_____ H _____ =   H   H   H

_____ C _____ =   H C O C O C H

_____ O _____ = _____   H   H   H

                                    =====

Rule 5

No. of electrons in structure _____

No. of valence electrons _____

Structure is _____

Unsaturated Groups

11. When the compound in question is **unsaturated**, the application of Rule 5 shows that the number of valence electrons in the trial structure is larger than the number of available valence electrons. In such cases the trial structure cannot be correct. It must be modified to contain the proper number of electrons. The skeleton of ethylene (C_2H_4) is

Each of __4__ hydrogen atoms will contribute _____ electron and each of _____ carbon atoms will contribute _____ electrons. The total number of available valence electrons is _____.

$$\underline{4} \quad H \quad \underline{1 \times 4 \; = \; 4}$$
$$\underline{2} \quad C \quad \underline{4 \times 2 \; = \; 8}$$
$$\underline{\underline{12}}$$

12. Adding electrons to the skeleton by making single bonds between all bonded atoms gives

Each hydrogen atom now has a pair of electrons, but each carbon has only __6__ electrons. Adding a pair of electrons to each carbon gives the trial structure

The number of electrons in the trial structure is _____. Since this exceeds the number of available valence electrons the structure is <u>incorrect</u>.

RULE 6

When the number of electrons in the trial structure is larger than the number of available valence electrons, the structure may be corrected by introducing one or more multiple bonds.

13. This is done by removing an unshared pair from each of two adjacent atoms and adding one electron pair as a second bond between the atoms. Each such operation reduces the number of electrons in the trial structure by two. Removing the unshared pairs of electrons on the carbon atoms and adding a second carbon–carbon bond gives

$$H{\diagdown}{\diagup}H$$
$$C{=}C$$
$$H{\diagup}{\diagdown}H$$

a structure in which there are four electrons involved in a **double bond** between the carbon atoms. The trial structure now contains _____ electrons and is correct.

14. Formaldehyde has the skeleton

$$H$$
$$C \quad O$$
$$H$$

Each of _____ hydrogen atoms will contribute _____ electron, the carbon atom, _____ electrons, and the oxygen, _____ electrons. The total number of valence electrons is _____.

_____   H   _____   =

_____   C   _____   =

_____   O   _____   =   _____

$$\overline{\overline{}}$$

Adding single bonds to the skeleton gives

_____

and providing the carbon atom and the oxygen atom with an octet of electrons gives

_____

The number of electrons in the trial structure is _____. The structure is _____ (correct, incorrect). Since the trial structure is incorrect, the pair of unshared electrons on the carbon atom and one of the pairs of unshared electrons on the oxygen atom are removed. Adding a second carbon–oxygen bond gives

_____

The trial structure now contains _____ electrons and is _____ (correct, incorrect).

15. The skeleton of acetonitrile is

Each of _____ hydrogens will contribute _____ electron, each of _____ carbons will contribute _____ electrons, and the nitrogen will contribute _____ electrons. The total number of valence electrons is _____.

_____   H   _____   =

_____   C   _____   =

_____   N   _____   =   _____

_____

Adding single bonds to the skeleton gives

_____

and providing the carbon and nitrogen atoms with octets gives

_____

The number of electrons in the trial structure is _____. The trial structure is _____ (correct, incorrect). Removing one unshared pair of electrons from carbon and one pair from nitrogen and adding a second carbon–nitrogen bond gives

_____

The structure now contains _____ electrons and is _____ (correct, incorrect).

Performing the same operation again gives

_____

The trial structure now contains _____ electrons and is _____ (correct, incorrect).

Thus, you continue to remove pairs of unshared electrons from adjacent atoms and add multiple bonds (add unsaturation) until the number of electrons in the trial structure is equal to the number of available valence electrons.

*   *   *   *   *

Sometimes, you must decide where a multiple bond is best added.

16. Formic acid has the skeleton

$$
\begin{array}{c}
\text{O} \\
\text{H C} \\
\text{O H}
\end{array}
$$

The number of available valence electrons is _____.

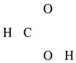

Filling in the skeleton with single bonds gives

$$
\text{H}-\text{C}\underset{\text{O}-\text{H}}{\overset{\text{O}}{\big<}}
$$

and adding the appropriate unshared pairs gives

$$
\text{H}-\overset{..}{\underset{..}{\text{C}}}\underset{\overset{..}{\underset{..}{\text{O}}}-\text{H}}{\overset{\overset{..}{\text{O}}{:}}{\big<}}
$$

The number of electrons in the structure is _____, which is __2__ too many electrons. This structure can be corrected by removing two unshared pairs and making one double bond. However, the double bond could be placed between the carbon atom and either one of the two oxygens. Thus,

$$H-C \underset{\overset{..}{\underset{..}{O}}-H}{\overset{O:}{\diagdown}} \qquad \text{or} \qquad H-C \underset{\overset{..}{\underset{..}{O}}-H}{\overset{:O:}{\diagup}}$$

The correct choice is the first structure and is made by observing that both oxygen atoms in the first structure have the appropriate valence of two. The second structure is less acceptable since it requires that the oxygen atoms have the unfamiliar valences of one and three respectively. We shall see later that the second structure is also less acceptable because it contains charge separation.

The elements most commonly found covalently bonded in organic compunds are listed below along with their valences.

carbon and silicon	4
hydrogen	1
oxygen and sulfur	2
the halogens (F, Cl, Br, I)	1
nitrogen and phosphorous	3 (4 in onium salts)
boron	3

One determines the valence of an atom in a structure by counting the number of electron pairs it is *sharing* with other atoms. That an atom may be sharing more than one pair with another atom (double or triple bond) does not change this. Thus, carbon may exhibit a valence of four in the following ways.

$$b-\overset{\overset{\textstyle a}{|}}{\underset{\underset{\textstyle c}{|}}{C}}-d \qquad \overset{a}{\underset{b}{\diagdown}}C{=}Z \qquad a-C{\equiv}Z$$

Oxygen may exhibit its valence of two as follows.

$$a-\overset{..}{\underset{..}{O}}-b \qquad :\overset{..}{O}{=}Z$$

Nitrogen may show a valence of three as follows.

$$a-\overset{..}{\underset{\underset{\textstyle b}{|}}{N}}-c \qquad a-\overset{..}{N}{=}Z \qquad \overset{..}{N}{\equiv}Z$$

17. The skeleton of acetyl chloride is

$$\begin{array}{ccc} H & & O \\ H & C & C \\ H & & Cl \end{array}$$

write the best Lewis structure for acetyl chloride by following the procedure in the previous problem.

Exceptions to the Octet Rule

There are stable compounds that do not obey the octet rule. A few of them are important players in organic chemistry.

Aluminum trichloride and boron trifluoride have only six electrons in the outer shell of the central atom.

Although stable under the right conditions, these "Lewis acids" are vigorously reactive toward compounds that have electrons to give. In these reactions they form new compounds in which the formerly electron-deficient atoms obtain an octet.

There are also many stable compounds in which at least one atom has <u>more</u> than an octet of electrons. The best Lewis structure for the sulfate ion ($SO_4^=$) is

The phosphate linkage in DNA is written

In each of these "hypervalent" structures the central atom has more than eight electrons. Hypervalence occurs in the elements of periods 3 to 6; note that these elements have atoms with low-lying d-orbitals that can accomodate additional electrons.

Exercises

Using the method outlined above, derive the structures for the following compounds. The unbonded skeletons are provided.

18. Propyne (C_3H_4)

Rule 3

_____ H _____ =

_____ C _____ = _____

Rule 4 (first trial)

H

H C C C H

H

Rule 5 (first trial)

No. of electrons in structure _____
No. of valence electrons _____
Structure is _____ (correct, incorrect)

(second trial)   **(third trial)**

_____ _____   _____ _____

Structure is _____   Structure is _____

19. Acetone (C_3H_6O)

Rule 3   **Rule 4**

H   O   H
H C   C   C H
H   H

Carry on.

20. Formamide (CH_3NO)

Rule 3   **Rule 4**

O

H   C   N   H

H

Carry on.

21. Urea (CH_4N_2O)

Rule 3   **Rule 4**

O

H   N   C   N   H

H   H

22. By now you have discovered that slavishly following rules 1–6 is tedious. As you see and use more and more Lewis structures, frequently you will be able to take shortcuts until writing correct Lewis structures becomes almost instinctive. See if you can arrive at the structure of benzene, C_6H_6, using 30 valence electrons and avoiding the trial structures.

H

H   C   C   H

C   C

C   C

H   H

H

Larger Molecules

So far, we have written the structures of small molecules. If the procedure outlined were applied scrupulously to writing the Lewis structures of larger molecules, the process would become tiresome. However, organic chemistry is largely the chemistry of functional groups. Often in writing mechanisms or other electron-pushing operations it is only necessary to write the Lewis structure of the functional group. The rest of the molecule can be stipulated in one of several shorthand notations that organic chemists are fond of using. Thus,

  becomes

and

  becomes

and

Most organic textbooks do not include the unshared pairs in functional groups. The structures above then become

The idea is to delete the unshared pairs and assume that even the beginning student will understand that they must be there because the octet rule applies. This is really a most unfortunate practice because so much of the chemistry of a functional group is determined by the presence (or absence) of unshared pairs.

In order to write Lewis structures of functional groups attached to any alkyl (R) or aryl (Ar) group, one writes down the skeleton (using some shorthand notation) and then counts the available valence electrons allowing each R or Ar attached directly to the functional group to bring one electron into the union.

   The use of R or Ar in organic structures seems to confuse some. The letter R is used to denote *any* alkyl (**aliphatic**) group, and Ar is used to denote *any* aryl (**aromatic**) group. This shorthand device allows a chemist to avoid writing a complicated organic structure when all that needs to be shown is the transformation of a functional group. For example,

$$R-CH_2-\overset{..}{\underset{..}{O}}-H \quad \xrightarrow{PBr_3} \quad R-CH_2-\overset{..}{\underset{..}{Br}}:$$

is much easier than

In this case R =

Likewise

$$Ar-\overset{\overset{\displaystyle \overset{..}{O}:}{\|}}{C}-\overset{..}{\underset{..}{O}}-H \quad \xrightarrow{SOCl_2} \quad Ar-\overset{\overset{\displaystyle \overset{..}{O}:}{\|}}{C}-\overset{..}{\underset{..}{Cl}}:$$

is easier than

In this case Ar =

23. Phenyl methyl ketone (acetophenone) has the skeleton

The number of available valence electrons is: from the phenyl group, __1__; from the methyl group, __1__; from the carbon atom, __4__; and from the oxygen atom __6__. The total number of valence electrons is _____. Filling in the skeleton with only single bonds gives

and adding the appropriate unshared pairs gives

The number of valence electrons in this structure is _____ which is _____ too many. Removing an unshared pair from carbon and another from oxygen and placing a second bond between carbon and oxygen gives the correct structure:

_____

24. The skeleton of benzyldimethylamine is

The number of available valence electrons is: from the benzyl group, _____; from each of two methyl groups, _____; and, from the nitrogen atom, _____, for a total of _____. Filling in the skeleton with single bonds gives

_____

and adding the appropriate unshared pair gives

The number of electrons in the functional group of this structure is _____, and the structure is _____ (correct, incorrect).

25. The skeleton is benzaldoxime is

The number of valence electrons is: from the phenyl group, _____, from each of two hydrogens, _____; from the carbon atom, ____; from the nitrogen atom, _____; and from the oxygen atom, _____, for a total of _____. Filling in the skeleton with single bonds and adding the appropriate unshared pairs gives

----

The number of electrons in the functional group of this trial structure is ____ which is ____ too many. Removing an unshared pair from carbon and nitrogen and adding a second bond between them gives

----

in which carbon, nitrogen, and oxygen have their customary valences of _____, _____, and _____ respectively. The alternative structure with a double bond between nitrogen and oxygen is

----

This structure is not acceptable because it requires carbon and oxygen to exhibit the unfamiliar valences of _____ and _____.

Exercises

Derive Lewis structures for the compounds below.

26. 2-phenyl-2-hexanol

$CH_3 - CH_2 - CH_2 - CH_2$

CH$_3$

C

O

H

27. Furan

H   H

C   C

H   C   C   H

O

28. Benzophenone phenylhydrazone

H

C   N   N

29. Azobenzene

    N   N

30. Methyl benzimidate

H

N

C   O   CH$_3$

31. Ethyl crotonate (For the purpose of drawing Lewis structures of compounds with more than one functional group, each group can be treated independently.)

$$
\begin{array}{cccc}
 & \text{H} & \text{H} & \text{O} \\
\text{CH}_3 & \text{C} & \text{C} & \text{C} \\
 & & & \text{O} \quad \text{CH}_2\!-\!\text{CH}_3
\end{array}
$$

Formal Charge

Some organic functional groups, although neutral overall, have formal charges on individual atoms. In addition, many of the intermediates that appear in organic reaction mechanisms are charged.

To calculate the **formal charge** on an atom in a particular structure it is necessary to distinguish between the electrons that make up an atom's octet and the electrons that formally "belong" to an atom. The distinction is an arbitrary one, but it is helpful in calculating formal charge. In any Lewis structure *all* electrons associated with an atom either as an unshared pair or in bonding to another atom (shared pair) are part of that atom's octet.

Chloromethane has the Lewis structure

$$
\begin{array}{c}
\text{H} \\
| \\
\text{H}-\text{C}-\ddot{\text{C}}\text{l}: \\
| \\
\text{H}
\end{array}
$$

Circling the carbon atom and its octet of electrons gives

Circling the chlorine atom and its octet gives

$$
\begin{array}{c}
\text{H} \\
| \\
\text{H}-\text{C}\!-\!\ddot{\text{C}}\text{l}: \\
| \\
\text{H}
\end{array}
$$

32. The Lewis structure of acetone is

$$
\begin{array}{c}
\ddot{\text{O}}: \\
\| \\
\text{CH}_3-\text{C}-\text{CH}_3
\end{array}
$$

Circling the carbonyl carbon, i.e., the carbon atom attached to oxygen, and its octet gives

$$
\begin{array}{c}
\ddot{\text{O}}: \\
\| \\
\underline{\text{CH}_3-\text{C}-\text{CH}_3}
\end{array}
$$

Circling the oxygen atom and its octet gives

$$\overset{\displaystyle \ddot{O}:}{\underset{\displaystyle CH_3-C-CH_3}{\|}}$$

Thus, atoms share electrons in making bonds, and a pair of electrons may be included in the octet of two different atoms.

When computing the formal charge on an atom, the number of electrons that "belong" to that atom is compared with the number of electrons the atom would have in the unbonded and neutral state. If the two numbers are the same, the formal charge on the atom is zero. In a Lewis structure *both* electrons in an unshared pair belong to the atom, and *one* of every pair of shared (bonding) electrons belongs to the atom.

33. Chloromethane has the Lewis structure

     _____

     The carbon atom is sharing __4__ electron pairs. In each shared pair the carbon atom "owns"
     __1__ electron. The number of electrons that "belong" to carbon is _____. Carbon, being a Group
     _____ element would have __4__ outer shell electrons in the unbonded, neutral state. Therefore,
     the carbon atom in chloromethane has a formal charge of <u>zero</u>.

34. In the Lewis structure for chloromethane, the chlorine atom is sharing _____ electron pair and
     "owns" _____ of those electrons. Also, the chlorine atom possesses two electrons from each of
     _____ unshared pairs. The total number of electrons that belong to chlorine is __7__ . Chlorine is
     a Group _____ element. The formal charge on chlorine in chloromethane is _____.

35. The Lewis structure for acetone is

$$\overset{\displaystyle \ddot{O}:}{\underset{\displaystyle CH_3-C-CH_3}{\|}}$$

     The carbonyl carbon is sharing _____ pairs of electrons (two carbon–carbon bonds and one
     carbon–oxygen double bond). From each of those shared pairs the carbonyl carbon "owns"
     _____ electron. The total number of electrons belonging to the carbonyl carbon is _____. Carbon
     is a Group _____ element. In acetone the formal charge on the carbonyl carbon is _____.

36. The oxygen atom in acetone possesses ____ unshared pairs and ____ shared pairs of electrons. The number of electrons that belong to oxygen is____. Oxygen is a Group ____ element. The formal charge on oxygen in acetone is _____.

*   *   *   *   *

All of the structures introduced in this chapter so far have only atoms with formal charge equal to zero. We will now see molecules containing atoms with formal charge other than zero.

37. Nitrobenzene has the skeleton

The number of available valence electrons is: from the phenyl group, _____; from the nitrogen atom, _____; and from each of two oxygen atoms, _____, for a total of _____. Filling in all single bonds and adding the appropriate unshared pairs gives

_____

The **functional group** of this structure contains _____ electrons. Therefore, unshared pairs are removed from the nitrogen atom and one of the oxygen atoms; a double bond is added giving

_____

which has the correct number of electrons. In this structure the nitrogen atom is sharing ____ pairs of electrons. From each shared pair the nitrogen owns ____ electron for a total of ____. Nitrogen is a Group ____ element and would have ____ outer shell electrons in the unbonded, neutral state. Since the nitrogen atom in nitrobenzene has one fewer electron than it would in the neutral state, it has a formal charge of __+1__. This is added to the Lewis structure as + giving

Since an electron is negatively charged, a shortage of one electron results in a single positive

charge (+) on an atom. Conversely, an excess of one electron results in a single negative charge (-) on an atom. When an atom in a Lewis structure "owns" two less electrons than it would have in the neutral, unbonded state, it is denoted by ++ or +2 and, conversely, = or -2.

38. Nitrobenzene, a neutral molecule, is not an ion. There must be a formal negative charge somewhere in the molecule to balance the positive charge on the nitrogen. The oxygen atom that is bonded to the nitrogen by a double bond

"owns" _____ unshared pairs of electrons. It is sharing _____ pairs of electrons. The number of electrons owned by the oxygen is _____. Since oxygen is a Group _____ element, the formal charge on this oxygen atom is _____. The oxygen atom that is bonded to the nitrogen by a single bond

owns _____ unshared pairs of electrons and is sharing _____ pair. The number of electrons owned by this oxygen is _____. This oxygen atom has one more electron than it would have in the neutral state and, thus, has a formal charge of _____. This is added to the Lewis structure as – giving

which is a complete, correct Lewis structure.

The cognoscenti in the crowd will recognize that this structure is one of four equivalent resonance structures for nitrobenzene. This will come up later in the text.

Exercises

Compute and add on the formal charges in these Lewis structures.

39. Pyridine N-oxide

40. Benzenesulfonic acid

41. N-methyl benzenesulfonamide

42. Ethylidenetriphenylphosphorane (an ylide)

43. Methylazide

44. Diazomethane

45. Phenylisocyanide

46. Phenylcyanide (benzonitrile)

47. Trimethylamine oxide

48. Dimethylsulfoxide

Ions

The structures encountered so far have been those of molecules, i.e., neutral species. In many cases there was no formal charge anywhere in the structure. In some cases formal charges appeared in the structure, but their algebraic sum was zero.

Although the starting materials and products of most organic reactions are molecules, a very large number of organic reactions involve ions as intermediates. Ions are charged species. The algebraic sum of the formal charges in an ion is not zero. With very few exceptions the ions encountered in organic chemistry will have a total charge of +1 or -1. Ions with a total charge of greater than one, which are quite common in inorganic chemistry, are rarely found in organic chemistry.

The ionic intermediates of organic chemistry are usually short-lived species because they are unstable relative to neutral molecules. Nevertheless, the proposed existence of ionic intermediates has contributed enormously to understanding organic reactions.

The most direct way to comprehend the structures of ions is to see how they arise in a reaction. This involves some sort of bond-breaking or bond-making process. Bond breaking and bond making are covered extensively in Chapter 3. They will be introduced briefly here.

Cations

Consider a molecule consisting of a methyl group attached to chlorine (chloromethane)

$$\begin{array}{c} H \\ | \\ H-C-\ddot{\underset{\cdot\cdot}{Cl}}: \\ | \\ H \end{array}$$

The formal charges on all the atoms in this molecule are zero. Now consider what would result if this molecule were broken at the C–Cl bond so that the two electrons that constitute the C–Cl bond both depart with the chlorine.

A Digression

We have a useful notation for the process. It uses curved arrows to indicate the movement of electrons. When organic chemists use the method, they are accused of "pushing electrons." The notation

$$
\begin{array}{c}
\text{H} \\
| \\
\text{H—C—Cl:} \\
| \\
\text{H}
\end{array}
$$

exactly illustrates chloromethane's being "broken at the C–Cl bond so that the two electrons that constitute the C–Cl bond both depart with the chlorine."

Carefully note these three points:

1. The base of the arrow begins *at the original location of the pair of electrons* (in this case, the C–Cl bond).

2. The head of the arrow points *to the destination* of the electrons (the chlorine atom).

3. When the head of the arrow has *two* barbs, it denotes the movement of a *pair* of electrons.

49. The usual method is used to calculate the formal charge on the resulting fragments. The carbon atom of the methyl fragment is sharing ____ electron pairs. In each shared pair the carbon "owns" _____ electron. The number of electrons which "belong" to carbon is _____. Carbon, being a Group _____ element would have _____ outer shell electrons in the neutral, unbonded state. Since the carbon atom has one fewer electron than it would in the neutral state it has a formal charge of _____. Therefore, the correct Lewis structure for the methyl cation is

$$
\begin{array}{c}
\text{H} \\
| \\
\text{H—C+} \\
| \\
\text{H}
\end{array}
$$

The chlorine fragment now possesses _____ unshared electron pairs. The number of electrons which belong to chlorine is _____. This Group VII element has a formal charge of _____ and has become the familiar chloride ion. Note the conservation of charge. In a chemistry equation the algebraic sum of the charges on one side *must* equal the sum on the other side. Thus, in this example, a neutral molecule (charge = 0) yields a cation plus an anion [charge =(+1) + (-1) = 0].

The arrow provides a strong indication of what formal charges will result. If a neutral molecule is cleaved, an excess of electrons (negative charge) will result at the head of the arrow and a deficiency of electrons (positive charge) will result at the tail of the arrow.

The reaction,

$$
\text{CH}_3\text{—Cl:} \quad \longrightarrow \quad \text{CH}_3^+ \quad + \quad :\!\text{Cl}\!:^-
$$

provides an example where the two electrons of the C–Cl bond both went with the chlorine. Two ions, one electron-poor and the other electron-rich, result from this **heterolytic** (unsymmetrical) bond **cleavage**.

50. The *n*-propyl cation can be formed from a molecule such as

$$CH_3-CH_2-CH_2-\overset{..}{\underset{..}{Cl}}:$$

When the C–Cl bond is broken so that both electrons leave with Cl, the fragments formed are

_____   + $:\overset{..}{\underset{..}{Cl}}:$

The carbon atom that had been attached to Cl is now sharing _____ electron pairs. In each shared pair the carbon atom owns _____ electron. The number of electrons which belong to carbon is _____. The formal charge on the carbon atom is _____. The correct Lewis structure for the *n*-propyl cation is

_____   + $:\overset{..}{\underset{..}{Cl}}:^-$

51. The isopropyl cation can be formed from

$$CH_3-\overset{\overset{\displaystyle H}{|}}{\underset{\underset{\displaystyle CH_3}{|}}{C}}-\overset{..}{\underset{..}{Br}}:$$

When the C–Br bond is broken so that both electrons go with Br (bromine departs as bromide ion), the organic fragment formed is

_____

The carbon atom that had been attached to bromine is now sharing _____ electron pairs. The total number of electrons that belong to carbon is _____. The formal charge on the carbon atom is _____. The correct Lewis structure for the isopropyl cation is

_____   +

52. The cyclopentyl cation can be formed from

When the iodine departs as iodide ion the organic fragment formed is

_____

The carbon atom that had been attached to iodine is now sharing _____ electron pairs. The total number of electrons belonging to the carbon atom is _____. The formal charge on the carbon atom is _____. The correct Lewis structure for the cyclopentyl cation is

_____   +   :Ï:⁻

53. Methanol, $CH_3-\overset{..}{\underset{..}{O}}-H$, is a compound in which the formal charge on all the atoms is zero. Consider what results when a proton, H⁺, becomes bonded to methanol by way of one of the unshared electron pairs on the oxygen atom, i.e.,

This time we use the curved arrow to signify bond making. Now a pair of unshared electrons on oxygen is pushed toward the region between the oxygen atom and the hydrogen ion. It becomes an O–H covalent bond.

In the resulting structure the oxygen atom owns one electron from each of _____ shared pairs and two electrons from _____ unshared pair. The total number of electrons which belong to the oxygen atom is ___. Oxygen is a Group ___ element. Since the number of electrons which the oxygen atom owns in this structure is one less than it would have in the neutral, unbonded state, the charge on oxygen is ___. The correct Lewis structure for the conjugate acid of methanol is

_____

Once again charge is conserved. Joining a neutral molecule and a cation *must* yield a cation. Again the formal charge distribution on the resulting ion is predictable from the arrow. Electrons are pushed *away* from the oxygen atom, leaving it with a positive charge. Electrons are pushed *toward* the hydrogen ion, neutralizing its erstwhile positive charge.

54. When a proton becomes bonded to diethyl ether, $CH_3-CH_2-\overset{\cdot\cdot}{\underset{\cdot\cdot}{O}}-CH_2-CH_3$, by way of one of the unshared electron pairs on the oxygen atom, the result is

$$CH_3-CH_2-\overset{\cdot\cdot}{\underset{\cdot\cdot}{O}}-CH_2-CH_3 \longrightarrow \underline{\hspace{3cm}}$$

with H^+

In this structure the oxygen atom owns one electron from each of ___ shared pairs and two electrons from ___ unshared pair. The total number of electrons that belong to oxygen is ___. The formal charge on oxygen is ___. The correct Lewis structure for the conjugate acid of diethyl ether is

_____

55. Tetrahydrofuran has the structure,

When a proton becomes bonded to tetrahydrofuran the result is

_____

In this structure the oxygen atom owns one electron from each of _____ shared pairs and two electrons from _____ unshared pair. The formal charge on the oxygen atom is _____. The Lewis structure for the conjugate acid of tetrahydrofuran is

_____

56. Methylamine has an unshared electron pair on the nitrogen atom. When a proton becomes bonded to methylamine, the result is

_____

In this structure the nitrogen atom is sharing _____ electron pairs. From each shared pair the nitrogen owns _____ electron for a total of _____ electrons. Nitrogen is a Group _____ element. The formal charge on the nitrogen atom is _____. The Lewis structure for the methylammonium ion is

_____

57. When a proton becomes bonded to diethylamine the result is

_____

In this structure the nitrogen owns one electron from each of _____ shared pairs. The total number

of electrons which belong to nitrogen is _____. The formal charge on the nitrogen is _____. The
Lewis structure for the diethylammonium ion is

_____

58. The structure of pyridine is

When a proton becomes bonded to the nitrogen atom by way of its unshared electron pair, the
result is

_____

In this structure the nitrogen atom is sharing _____ electron pairs. The formal charge on the
nitrogen atom is _____. The Lewis structure for the pyridinium ion is

_____

Anions

In methyl lithium the carbon atom is bound to a metal atom. The formal charges on all the atoms in
the molecule are zero. Consider the result if the C–Li bond were broken so that the two electrons that
constitute the bond both *remain* with the carbon.

59. The carbon atom owns one electron from each of _____ shared pairs and two electrons from

_____ unshared pair. The number of electrons that belong to carbon is _____. Carbon is a Group _____ element. Since the carbon atom has one more electron than it would in the neutral, unbonded state, it has a formal charge of __-1__. The Lewis structure for the methyl anion is

$$
\begin{array}{c}
H \\
| \\
H-C\!:\;^- \\
| \\
H
\end{array}
$$

The lithium fragment must have a formal charge of _____.

60. The *n*-butyl anion can be formed from

$$
CH_3-CH_2-CH_2-\overset{\displaystyle H}{\underset{\displaystyle H}{\overset{|}{\underset{|}{C}}}}-Li
$$

When the C–Li bond is broken so that both electrons remain with the carbon atom, the organic fragment formed is

_____

The carbon atom that had been attached to Li now owns one electron from each of _____ shared pairs and two electrons from _____ unshared pair. The number of electrons that belong to carbon is _____. The formal charge on the carbon atom is _____. The Lewis structure for the *n*-butyl anion is

$$
+\;\; Li^+
$$

_____

61. The isobutyl anion can be formed from

$$
CH_3-\overset{\displaystyle H}{\underset{\displaystyle CH_3}{\overset{|}{\underset{|}{CH}}}}-\overset{\displaystyle H}{\underset{\displaystyle H}{\overset{|}{\underset{|}{C}}}}-Na
$$

When the C–Na bond is broken so that both electrons remain with the carbon atom, the organic fragment formed is

_____

The carbon atom that had been attached to Na now owns one electron from each of _____ shared pairs and two electrons from _____ unshared pair. The number of electrons that belong to carbon is _____. The Lewis structure for the isobutyl anion is

$$+ \quad Na^+$$

_____

62. When the C–Mg bond in

is broken so that both electrons remain with the carbon atom, the fragment formed is

_____

The carbon atom that had been attached to Mg now owns one electron from each of _____ shared pairs and two electrons from _____ unshared pair. The formal charge on the carbon atom is _____. The Lewis structure for the cyclohexylmethyl anion is

$$+ \quad ^+MgBr$$

_____

The main player in organic chemistry is carbon. It follows that organic chemists concentrate on the fate of carbon in these reactions. Compare the cleavage reactions (59–62) with (49–52). All are heterolytic cleavages. In 59–62 the electrons move toward the carbon, generating a **carbanion.** Conversely, in 49–52 the electrons move away from carbon and produce a **carbocation.** The original polarity of the bond influences the direction of heterolysis; the more electronegative atom takes the electrons. Thus, the electrons go with the carbon when a carbon–metal bond breaks, but go with the halogen when a carbon–halogen bond breaks.

63. Ethanol, $CH_3—CH_2—\overset{..}{O}—H$, is a compound in which the formal charge on all the atoms is zero. Under certain conditions the O–H bond can be broken so that both electrons remain with the oxygen atom. The resulting fragments are

$$CH_3—CH_2—\overset{..}{\underset{..}{O}}—H \longrightarrow CH_3—CH_2—\overset{..}{\underset{..}{O}}: \ + \ H$$

In this structure the oxygen owns one electron from _____ shared pair and two electrons from each of _____ unshared pairs. The total number of electrons belonging to oxygen is _____. Oxygen is a Group _____ element. The formal charge on the oxygen atom is _____. The correct Lewis structure for the ethoxide ion is

$$+ \ \ H^+$$

————————————

Note that the other fragment, the proton, leaves with a formal charge of +1.

64. The loss of a proton attached to the oxygen atom of *t*-butyl alcohol results in

$$+ \ \ H$$

————————————

In this structure the oxygen atom owns one electron from _____ shared pair and two electrons from each of _____ unshared pairs. The number of electrons belonging to oxygen is _____. The formal charge on the oxygen is _____. The Lewis structure for the *t*-butoxide ion is

$$+ \ \ H^+$$

————————————

65. A very strong base can remove a proton from methylamine,

$$CH_3 - \overset{\displaystyle ..}{N} \overset{\displaystyle \diagup H}{\diagdown H}$$

The proton removed is one of those attached to the nitrogen atom. The resulting fragments are

+   H

_____

The nitrogen atom now owns one electron from each of _____ shared pairs and two electrons from each of _____ unshared pairs. The number of electrons that belong to the nitrogen atom is _____. Nitrogen is a Group _____ element. The formal charge on nitrogen is _____. The Lewis structure for the methylamide ion is

+ H^+

_____

66. The loss of a proton from diisopropylamine,

$$CH_3 - CH - \overset{\displaystyle ..}{\underset{\displaystyle |}{N}} - CH - CH_3$$

with CH_3 groups on the CH carbons and H below the N

results in the fragment

_____

The number of electrons which belong to the nitrogen atom is _____. The formal charge on the nitrogen atom is _____. The correct Lewis structure for the diisopropylamide ion is

+ H^+

_____

Free Radicals

Covalent bonds can suffer *homolytic* (symmetrical) bond breaking. During homolytic cleavage, the electrons of a covalent bond become unpaired and one electron goes with each fragment. Radicals, rather than ions, are formed. Radicals usually have a formal charge of zero.

Consider the homolytic cleavage of a Cl_2 molecule.

The entirely symmetrical process yields two chlorine atoms. Since each chlorine "owns" seven electrons, the formal charge is zero. The presence of an *unpaired* electron identifies each fragment as a "free radical."

Note carefully how the curved arrow notation has changed! The base of the arrow still begins at the original location of the electron and the arrow's head still points to the destination of the electron. However, now the head of the arrow has *only one barb* signifying the movement of a single electron. Two such arrows are required to denote homolytic bond breaking since each electron goes its separate way.

67. When a Br_2 molecule undergoes homolytic cleavage, it yields _____ (number) bromine radicals. Thus,

Each radical has a formal charge of zero.

Homolytic bond breaking usually requires quite large energies. The homolysis of Cl_2 or Br_2 is induced only at high temperature or by ultraviolet light. Many free radical reactions are promoted photochemically. Homolytic cleavage reactions important to organic chemistry often occur when the covalent bond in question is weak. Oxygen–oxygen (peroxide) bonds are a good example.

68. When di-*tert*-butyl peroxide breaks homolytically, it forms two identical radicals with the Lewis structure

The oxygen atom owns _____ electron from a shared pair, _____ unshared electrons for a total

of _____ electrons. Oxygen is a Group VI element, and its formal charge is _____.

69. The homolysis of the O–O bond in diacetyl peroxide gives two acetyl radicals with Lewis structure

The formal charge on this radical is _____.

70. Homolysis of an unsymmetrical bond will result in two different radicals. The homolytic cleavage of the O–O bond in *tert*-butyl hydroperoxide will yield

71. Homolysis of *m*-chloroperbenzoic acid will give

72. Carbon radicals are usually produced indirectly. An exception is the homolytic cleavage of azobis(isobutyronitrile). In this commercially important reaction, the C–N bonds break homolytically to give dinitrogen plus two identical carbon radicals.

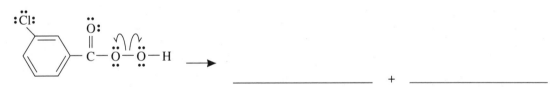

Reaction conditions dictate how molecules cleave. Write the products from isopropyl hydroperoxide in each of the reactions described.

73. *Hetero*lytic cleavage of the O–H bond to release a proton.

$\longrightarrow$ _____   +   H⁺

74. *Homo*lytic cleavage of the O–O bond.

$\longrightarrow$ _____   +   _____

75. *Hetero*lytic cleavage of the C–O bond to yield a carbocation.

$\longrightarrow$ _____   +   _____

Answers

Chapter 1

1. IV, 4, 6

2. 1, 4, 6, 14

      4   H $1 \times 4 =$ 4

      1   C $4 \times 1 =$ 4

      1   O $6 \times 1 =$ <u>6</u>

                           14

3.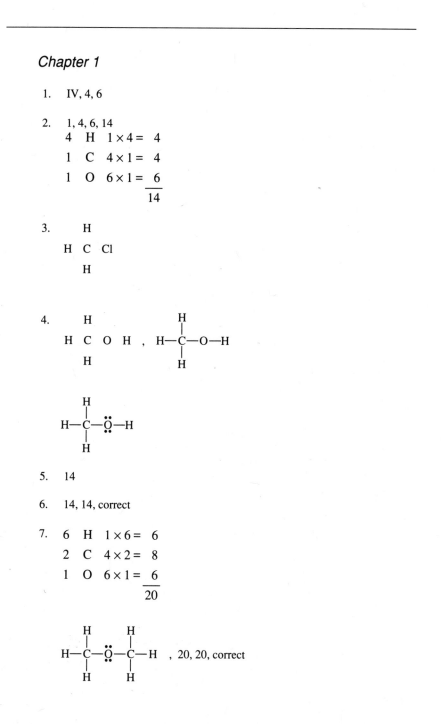

4.

5. 14

6. 14, 14, correct

7. 6 H $1 \times 6 =$ 6

    2 C $4 \times 2 =$ 8

    1 O $6 \times 1 =$ <u>6</u>

                         20

   , 20, 20, correct

8. 5 H $1 \times 5 =$ 5   ,

    1 C $4 \times 1 =$ 4

    1 N $5 \times 1 =$ 5

    ————

    14

, 14, 14, correct

9. 4 H $1 \times 4 =$ 4   ,

    1 C $4 \times 1 =$ 4

    1 S $6 \times 1 =$ 6

    ————

    14

, 14, 14, correct

H—C—S—H (structure)

10. 8 H $1 \times 8 =$ 8   ,

    3 C $4 \times 3 =$ 12

    2 O $6 \times 2 =$ 12

    ————

    32

, 32, 32, correct

H—C—O—C—O—C—H (structure)

11. 1, 2, 4, 12

12. 14

13. 12

14. 2, 1, 4, 6, 12,   2 H $1 \times 2 =$ 2   ,

    1 C $4 \times 1 =$ 4

    1 O $6 \times 1 =$ 6

    ————

    12

(two structures: C—O, and C—Ö:)

14, incorrect,   (structure) C=Ö: , 12, correct

15. 3, 1, 2, 4, 5, 16,   3 H $1 \times 3 =$ 3   ,

    2 C $4 \times 2 =$ 8

    1 N $5 \times 1 =$ 5

    ————

    16

H—C—C—N, (structure)

H—C—C—N: , 20, incorrect,   H—C—C=N: , 18, incorrect,

H—C—C≡N: , 16, correct,

16. 18,   2 H $1 \times 2 =$ 2   ,   20

    1 C $4 \times 1 =$ 4

    2 O $6 \times 2 =$ 12

    ————

    18

17.

18.   4 H $1 \times 4 = 4$,

 3 C $4 \times 3 = \underline{12}$

     16

19.   6 H $1 \times 6 = 6$

 3 C $4 \times 3 = 12$

 1 O $6 \times 1 = \underline{6}$

     24

20.

 3 H $1 \times 3 = 3$,

 1 C $4 \times 1 = 4$

 1 O $6 \times 1 = 6$

 1 N $5 \times 1 = \underline{5}$

     18

21. 4 H $1 \times 4 =$ 4   ,   :Ö: ,   26
     1 C $4 \times 1 =$ 4   24
     1 O $6 \times 1 =$ 6   H—N̈—C—N̈—H   incorrect
     2 N $5 \times 2 = \underline{10}$ | |
                              24   H   H

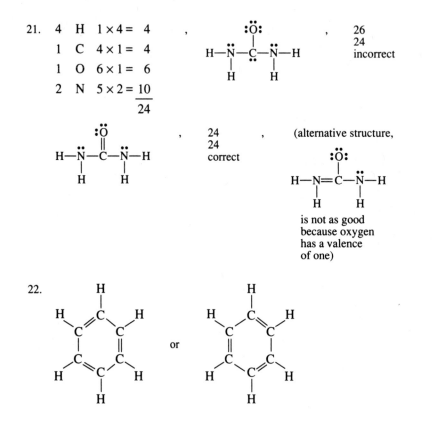

        :Ö   ,   24   ,   (alternative structure,
        ‖   24
   H—N̈—C—N̈—H   correct   :Ö:
        | | |
        H   H   H—N̈=C—N̈—H
                                                             | |
                                                             H   H

                                                    is not as good
                                                    because oxygen
                                                    has a valence
                                                    of one)

22.

or

23.   12, 14, 2,

24.   1, 1, 5, 8,

                           , 8, correct

25.   1, 1, 4, 5, 6, 18,   , 20, 2,

, 4, 3, 2,   , 3, 3

26.

27.

28.

29.

30.

31.

32.

33.

34.   1, 1, 3, VII, zero

35.   4, 1, 4, IV, zero

36.   2, 2, 6, VI, zero

37.   1, 5, 6, 18, , 4, 1, 4, V, 5

38.   2, 2, 6, VI, zero, 3, 1, 7, -1

39.

40.

41.

42.

43.

44.

45.

46. , (formal charge is zero on all atoms)

47.

$$CH_3 \overset{\overset{\displaystyle CH_3}{|}}{\underset{\underset{\displaystyle CH_3}{|}}{N}}^+ \!\!\!\!- \overset{\cdot\cdot}{\underset{\cdot\cdot}{O}}\!\!:^-$$

48.

$$\overset{:\overset{\cdot\cdot}{O}:^-}{|}$$
$$CH_3 \underset{+\,\cdot\cdot}{-}S - CH_3$$

49. 3, 1, 3, IV, 4, +1, 4, 8, -1

50. $CH_3 - CH_2 - CH_2$, 3, 1, 3, +1, $CH_3 - CH_2 - CH_2^+$

51.

$$CH_3 \overset{\overset{\displaystyle H}{|}}{\underset{\underset{\displaystyle CH_3}{|}}{C}} \, , \quad 3, 3, +1, \; CH_3 \overset{\overset{\displaystyle H}{|}}{\underset{\underset{\displaystyle CH_3}{|}}{C}}^+$$

52.

$\langle$pentagon$\rangle$ — H , 3, 3, +1, $\langle$pentagon$\rangle^+_{+}$— H

53. 3, 1, 5, VI, +1, $CH_3 - \overset{\cdot\cdot}{O}{}^+ - H$
with H below

$$CH_3 - \overset{\overset{\displaystyle\cdot\cdot\,+}{}}{\underset{\underset{\displaystyle H}{|}}{O}} - H$$

54. $CH_3-CH_2-\overset{..}{\underset{\underset{H}{|}}{O}}-CH_2-CH_3$, 3, 1, 5, +1, $CH_3-CH_2-\overset{+\,..}{\underset{\underset{H}{|}}{O}}-CH_2-CH_3$

55.

, 3, 1, +1,

56. $\underset{\underset{H}{|}}{\overset{\overset{H}{|}}{CH_3-N-H}}$, 4, 1, 4, V, +1, $\underset{\underset{H}{|}}{\overset{\overset{+}{\overset{H}{|}}}{CH_3-N-H}}$

57. $\underset{CH_2-CH_2}{\overset{CH_3-CH_2}{\diagdown}}N\overset{\diagup H}{\diagdown H}$, 4, 4, +1, $\underset{CH_3-CH_2}{\overset{CH_3-CH_2}{\diagdown}}\overset{+}{N}\overset{\diagup H}{\diagdown H}$

58. ,4, +1,

59. 3, 1, 5, IV, +1

60. $\underset{\underset{H}{|}}{\overset{\overset{H}{|}}{CH_3-CH_2-CH_2-C:}}$, 3, 1, 5, –1, $\underset{\underset{H}{|}}{\overset{\overset{H}{|}}{CH_3-CH_2-CH_2-\overset{-}{C:}}}$

61. $\underset{\underset{CH_3}{|}\,\underset{H}{|}}{\overset{\overset{H}{|}}{CH_3-CH-C:}}$, 3, 1, 5, $\underset{\underset{CH_3}{|}\,\underset{H}{|}}{\overset{\overset{H}{|}}{CH_3-CH-\overset{-}{C:}}}$

62. ,3, 1, –1,

63. 1, 3, 7, VI, –1, $CH_3-CH_2-\overset{..}{\underset{..}{O}}:^-$

64.

$$CH_3-\underset{\underset{CH_3}{|}}{\overset{\overset{CH_3}{|}}{C}}-\overset{..}{\underset{..}{O}}: \quad , \quad 1,3,7,-1, \quad CH_3-\underset{\underset{CH_3}{|}}{\overset{\overset{CH_3}{|}}{C}}-\overset{..}{\underset{..}{O}}:^{-}$$

65. $CH_3-\overset{..}{\underset{..}{N}}-H$,   2, 2, 6, V, –1, $CH_3-\overset{..}{N}\overset{-}{=}H$

66.

$$CH_3-CH-\overset{..}{\underset{..}{N}}-CH-CH_3 \quad , \quad 6,-1, \quad CH_3-CH-\overset{-}{\underset{..}{N}}-CH-CH_3$$

with CH_3 groups above the CH positions on both structures.

67. 2, $:\overset{..}{\underset{..}{Br}}\cdot$, $\cdot\overset{..}{\underset{..}{Br}}:$

68.

$$CH_3-\underset{\underset{CH_3}{|}}{\overset{\overset{CH_3}{|}}{C}}-\overset{..}{\underset{..}{O}}\cdot \quad , 1\ 5, 6, \text{zero}$$

69.

$$CH_3-\overset{\overset{..}{\overset{\parallel}{O}}}{C}-\overset{..}{\underset{..}{O}}\cdot \quad , \text{zero}$$

70.

$$CH_3-\underset{\underset{CH_3}{|}}{\overset{\overset{CH_3}{|}}{C}}-\overset{..}{\underset{.}{O}}\cdot$$

71. $:\overset{..}{\underset{..}{Cl}}$ benzene ring with $-\overset{\overset{..}{\overset{\parallel}{O}}}{C}-\overset{..}{\underset{..}{O}}\cdot$, $\cdot\overset{..}{\underset{..}{O}}-H$

72.

$$CH_3-\underset{\underset{CH_3}{|}}{\overset{\overset{CN}{|}}{C}}\cdot$$

73. $\underset{CH_3}{\overset{CH_3}{>}}C\overset{H}{\diagup}$ with $\overset{..}{\underset{..}{O}}-\overset{..}{\underset{..}{O}}:^{-}$

74.

75.

$$CH_3$$
$$\diagdown$$ $$H$$
$$C\ +$$
$$\diagup$$
$$CH_3$$

, $$:\!\overset{..}{\overset{-}{O}}\!-\!\overset{..}{\overset{..}{O}}\!-\!H$$

Resonance Structures

2

Correct Lewis structures are indispensable to understanding the chemical and physical properties of organic molecules. As you will see in Chapter 3 of this text, they are essential to writing the mechanisms of organic reactions. As helpful as they are, however, single Lewis structures do not always describe an organic compound accurately.

By this time you should be acquainted with the resonance method. The resonance method, while still preserving the useful features of the Lewis structure, is a procedure used to describe accurately a molecule for which a single Lewis structure is not sufficient. The resonance method attempts to describe the true structure of a molecule as a hybrid of several Lewis structures. These Lewis structures are called *resonance structures*. It is *essential* to understand that there is only *one* structure for the actual molecule. That structure is not written down but takes its character from the sum of the resonance structures. Resonance structures have no discrete existence of their own but, taken in combination, describe the true structure of the molecule.

The resonance method consists of writing down all the possible resonance structures and making a judgment about which of the resonance structures will make important contributions to the hybrid. In order to do that, one must be able to write resonance structures.

There are several rules to which acceptable resonance structures for a compound must conform. First, the relative positions of all atoms in all resonance structures for a single compound must be the same. Second, all resonance structures for a single compound must have the same number of paired and unpaired electrons. It follows from this rule that the algebraic sum of formal charge on each resonance structure for a single compound must be the same. Third, the important resonance structures will have comparable energies.

That is, any resonance structure that has an energy substantially higher than the rest must be discarded. The energy of a particular structure depends on several things (charge distribution, the octet rule, **steric strain,** and so on). Judging the relative energies of a series of structures requires an intuitive sense that you develop through experience. This chapter concentrates solely on how to push electrons to generate a series of resonance structures and leaves the development of judgment to you, your professor, and the main text.

While these rules are important, they are not helpful to the student faced with the task of writing resonance structures. This chapter will present a method for writing resonance structures, starting with one Lewis structure and "pushing electrons" to generate new resonance structures. To do that one must find the proper combination of pushable electrons and places to which electrons can be pushed (receptors). In Lewis structures the pushable electron pairs are the unshared electron pairs on a single atom and the pi electron pairs in multiple bonds. Receptors can be (1) atoms with a formal positive charge, (2) atoms that can tolerate a formal negative charge, and (3) atoms which possess pushable electrons themselves.

Simple Ions

Cations

One Lewis structure for the allyl cation is

This structure contains a pair of pushable electrons, namely the **pi electrons** in the double bond between C_2 and C_3. The structure also contains a positively charged carbon atom at C_1 that can act as a receptor. A new resonance structure can be generated by pushing the pair of pi electrons to the receptor. This is signified by the curved arrow with two barbs introduced in Chapter 1.

By pushing electrons to the receptor, the positive charge has been neutralized, but a new positive charge has been generated at C_3. Structure 1 can be regenerated from (2) by pushing the pi electrons between C_1 and C_2 toward the positively charged carbon C_3. Structures (1) and (2) are equivalent but not identical. In the resonance notation this is written

$$\left[H_2C = CH - \overset{+}{C}H_2 \quad \longleftrightarrow \quad H_2\overset{+}{C} - CH = CH_2 \right]$$

Check to be sure that these two resonance structures obey the rules set out in the beginning of this chapter. Also, check the assignment of formal charge.

1. One Lewis structure for the 2-butenyl cation is $CH_3 - CH = CH - \overset{+}{C}H_2$. A new resonance structure can be generated by pushing the pi electrons to the receptor.

$$\underline{\qquad CH_3 - CH = CH - \overset{+}{C}H_2 \qquad} \qquad \text{generates} \qquad \underline{\qquad\qquad\qquad}$$

supply arrow

2. A second resonance structure for the 3-cyclopentenyl cation can be generated.

gives

_____

supply arrow   _____

3. One structure for the conjugate acid of acetone is

The _____ electrons in the carbon–oxygen double bond are pushable electrons, and the _____ atom is a receptor.

$$\underset{CH_3}{\overset{CH_3}{>}}C=\overset{+}{\underset{..}{O}}-H$$ generates

_____

4. Similarly, a resonance structure for the conjugate acid of 2-butanone can be written. Thus,

$$\overset{+\overset{..}{O}-H}{\underset{\|}{CH_3-C-CH_2-CH_3}}$$ generates

_____

supply arrow   _____

It is crucial that these arrows indicate precisely *what* electrons are being pushed *where*. Therefore, after doing these exercises check the answers carefully to see that you have inserted the arrows properly.

5. A second resonance structure for protonated cycloheptanone is produced by

giving

_____

supply arrow   _____

6. Pairs of unshared electrons can be pushed. One Lewis structure for the methoxymethyl cation is $CH_3-\overset{\cdot\cdot}{\underset{\cdot\cdot}{O}}-CH_2^+$. The structure contains a pair of pushable electrons, namely the unshared electrons on the _____ atom. The structure also contains a positively charged _____ atom that can act as a _____. A second resonance structure can be generated by pushing the unshared electrons to the receptor. Thus,

$$CH_3-\overset{\cdot\cdot}{\underset{\cdot\cdot}{O}}-\overset{+}{C}H_2 \qquad \text{generates}$$

_____   _____
        supply arrow

It is not possible to push electrons toward the other carbon, because it is not a receptor.

7. One structure for the acetoxonium ion is

$$CH_3-\overset{+}{C}=\overset{\cdot\cdot}{O}:$$
$$\textbf{(3)}$$

Clearly, the receptor is the positively charged _____ atom. However, there are available two different kinds of pushable electrons, namely, pi electrons in the carbon–oxygen double bond or unshared electrons on the oxygen atom. The proper choice in this case is dictated by the following considerations. In structure (3) the carbon atom possesses a formal positive charge and also has only six electrons in its outer shell (it lacks a stable octet of electrons). If the pi electrons are pushed, namely,

$$\textbf{(4)}$$

structure (4) is obtained in which the oxygen possesses a formal positive charge but, in addition, both the oxygen *and* the carbon atom lack a stable octet. This structure will be very unstable. Structure (4), despite the fact that it can be generated by properly pushing electrons, is not included in the resonance hybrid for the acetoxonium ion. If, on the other hand, the unshared electrons on the oxygen are pushed, a more acceptable structure is obtained, namely,

 $\longleftrightarrow$ _____

$$\textbf{(5)}$$

In structure (5) the oxygen atom possesses a formal positive charge, and both the carbon and the oxygen atom have a stable octet. Structure (5) is included in the resonance hybrid.

8. Another ion of this type is

$$CH_3-CH-\overset{+}{C}=\overset{..}{O}:$$
$$\quad\;\;\; |$$
$$\quad\;\; CH_3$$

_____

supply arrow

which gives the resonance structure

_____

*   *   *   *   *

It is important to recognize opportunities to push electrons and generate new resonance structures. It is also important to recognize those structures in which one *cannot* push electrons. In order to be able to generate one or more resonance structure, a Lewis structure must have pushable electrons and a receptor. Furthermore, the receptor must be *next to* the pushable electrons.

9. There are no important resonance structures for the isopropyl cation

$$CH_3-\overset{+}{CH}-CH_3$$

because there are no _____ _____ in the structure. There are no important resonance structures for dimethyl ether

$$CH_3-\overset{..}{\underset{..}{O}}-CH_3$$

because, although there are pushable electrons on the oxygen atom, there is no _____. There are no important resonance structures for the 5-pentenyl cation

$$H_2C=CH-CH_2-CH_2-\overset{+}{CH_2}$$

because the pushable electrons and the receptor are separated by two methylene groups. Thus, the electrons have no way to get to the receptor.

Exercises

In this exercise one Lewis structure is provided. Where possible, push electrons to generate another resonance structure. In some cases no additional structure can be generated.

10.

11.

12.

13.

14.

15.

$$CH_3-\overset{\overset{\displaystyle CH_3}{|}}{\underset{\underset{\displaystyle CH_3}{|}}{C}}-\overset{+}{C}H-\overset{\cdot\cdot}{\underset{\cdot\cdot}{O}}-CH_2-CH_3 \longleftrightarrow$$

16.

17.

$$\overset{+}{O}\equiv C-\bigcirc \longleftrightarrow$$

Anions

One Lewis structure for the acetate ion is

This structure contains several pairs of pushable electrons. Since the object of writing resonance structures is to show delocalization of charge, it is reasonable that electrons should be pushed *away* from a center of negative charge. Pushing a pair of unshared electrons, located on the negatively charged oxygen, toward the carbonyl carbon would give

leading to

This, however, gives the intolerable situation of having ten valence electrons around the carboxyl carbon. The situation can be relieved by pushing the pi electrons of the original carbon–oxygen double bond toward the oxygen atom, which can tolerate the negative charge. Thus,

gives

18. The cyclohexane carboxylate anion has a Lewis structure

Pushing a pair of unshared electrons away from the negatively charged oxygen atom and, at the same time, pushing a pair of pi electrons toward the other oxygen will generate a second resonance structure. Thus,

supply arrows

19. One Lewis structure for the enolate anion of acetaldehyde is

Pushing the pair of unshared electrons on the carbon atom away from the center of negative charge and pushing the pi electrons of the carbon–oxygen double bond to the oxygen atom generates a second resonance structure. Thus,

                    gives

_____   _____
        supply arrows

20. The allyl anion is a resonance hybrid. One of the two equivalent resonance structures is

$$\overset{-\,\,\cdot\cdot}{CH_2}- CH = CH_2$$

Delocalization of the unshared electrons on the negatively charged carbon atom and simultaneous delocalization of the pi electrons to the other terminal carbon atom generates another structure.

_____   ⟷   _____
    $\overset{-\,\,\cdot\cdot}{CH_2}- CH = CH_2$
        supply arrows

21. The acetonitrile anion provides a slight variation on this theme. Thus, one pair of pi electrons in a triple bond is pushed.

_____   ⟷   _____
    $\overset{-\,\,\cdot\cdot}{CH_2}- C \equiv N{:}$
        supply arrows

*   *   *   *   *

Here are several errors and misconceptions that appear frequently in attempts to write resonance structures.

1. Always push *electrons*, never push positive charges. Remember that the arrows used in generating resonance structures indicate how the electrons are moving. A structure such as

$$CH_2 = CH - \overset{+}{CH_2}$$

is incorrect.

2. Always push electrons away from the centers of negative charge and toward centers of positive charge. The resonance method depends on delocalizing charge. Electrons, which have negative charge, must be pushed away from centers of relatively high electron density and toward centers of relatively low electron density.

Exercises

In these exercises one Lewis structure is provided. Where possible, push electrons to generate another resonance structure. In some cases there are no additional resonance structures.

22.

23.

24. $:N{\equiv}C-\overset{..}{\underset{}{C}}H-CH_2-CH{=}CH_2$ ⟷

25.

26.

27.

28.

Free Radicals

Some free radicals are stabilized by electron delocalization. The resonance method depicts this well. The formalism is a bit different from that used for ions and you must pay close attention at first.

Electrons are pushed as pairs (arrows with two barbs) when writing resonance structures for molecules or ions. For free radicals, electrons are pushed individually (arrows with one barb) and some electrons become unpaired as others become paired.

The allyl radical has an unpaired electron and a pair of pi electrons located next to each other. The generation of an equivalent, but not identical, resonance structure is shown as the pi-pair becoming unpaired (a homolytic cleavage of the **pi bond**) and one of those electrons becoming paired with the erstwhile unpaired electron.

$$CH_2\!=\!CH\!-\!\overset{\textstyle\cdot}{C}H_2 \qquad\longleftrightarrow\qquad \overset{\textstyle\cdot}{C}H_2\!-\!CH\!=\!CH_2$$

The notation is cluttered with more arrows since each electron has its own arrow.

29. Draw the second resonance structure.

$$CH_3\!-\!CH\!=\!CH\!-\!\overset{\textstyle\cdot}{C}H_2 \qquad\longleftrightarrow$$
$$4\quad\ 3\quad\ 2\quad\ \ 1$$

_____

Note that no delocalization occurs at C_4.

30. Draw another resonance structure.

$$\begin{matrix} CH_3 \\ \\ CH_3 \end{matrix}\!\!\!\!\searrow\!\!\!\!\overset{\textstyle\cdot}{C}\!-\!CH\!=\!CH\!-\!CH_3 \qquad\longleftrightarrow$$

_____   _____

supply arrows

31. Resonance occurs in cyclic radicals.

32. Again.

supply arrows

33. Electrons can be delocalized from any multiple bond next to a radical center.

$$:N\equiv C-\overset{\displaystyle \cdot}{C}-CH_3$$
$$\underset{CH_3}{|}$$

⟷

_____   _____
supply arrows

34. Again.

CH$_3$ ... C—C ... O ...
H   CH$_3$

⟷

_____   _____
supply arrows

Exercises

One Lewis structure is provided along with the number of possible structures (including the one provided). Draw the other structures.

35. (3)   :O   N   O:

36. (3)

Benzene and Benzenoid Aromatic Compounds

The best known example of a resonance hybrid is benzene. Benzene has a Lewis structure

(6)

This structure contains three pairs of pushable electrons. It is expedient, however, to consider only one pair, say the pair of pi electrons between C_1 and C_2, as the pushable electrons and the others as receptors.

37. Drawing an arrow pushing the C_1–C_2 pi electrons toward C_3 would result in the unacceptable situation of having ten electrons around C_3. The situation is relieved by pushing the C_3–C_4 pi electrons toward C_5 giving

This presents C_5 with the same dilemma that C_3 had a moment ago. By pushing the C_5–C_6 pi electrons toward C_1, the dilemma is solved since C_1 is missing the pair of pushable pi electrons that were used to begin this process. Thus,

gives

_____

(6)   **(7)**

Structures (6) and (7) are referred to as Kekulé structures. Whenever a benzene ring appears in a compound, the two Kekulé structures should be the first two resonance structures written.

August Kekulé (1829–96) proposed and proved this structure for benzene in 1865. He recognized at the time that a single structure did not account for certain properties of benzene and proposed, not too convincingly, a dynamic equilibrium between the two equivalent but not identical structures. The resonance theory (1935) resolved this problem.

Several shorthand notations that imply the two Kekulé structures are used by organic chemists. They are

Each of these can be used to indicate the phenyl group. Thus, toluene can be written

or

38. A second resonance structure for *ortho*-xylene can be generated.

_____   _____

        supply arrows

39. When more than one benzene ring is present in the same molecule, the number of possible resonance structures is increased. In diphenylmethanol, a second resonance structure can be generated by pushing electrons in the ring on the right and leaving the ring on the left undisturbed.

_____

Conversely, pushing electrons in the ring on the left and leaving the right undisturbed gives another structure.

_____   _____

supply arrows

Finally, pushing electrons on both rings gives a fourth resonance structure.

_____   _____

supply arrows

The complete resonance notation for diphenylmethanol is

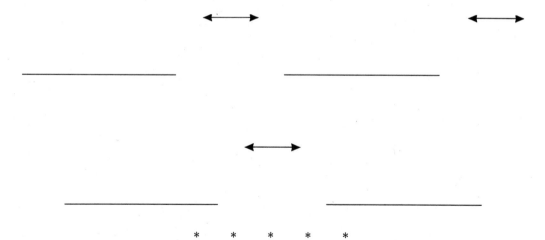

_____   _____   _____

*   *   *   *   *

"Fused" aromatic compounds contain benzene rings that have two carbon atoms in common. Writing resonance structures for fused aromatic compounds is done in a way similar to that described for diphenylmethanol. That is, electrons in one ring are pushed and the rest of the molecule is left undisturbed.

40. One Lewis structure for naphthalene is

Pushing electrons in the right ring and leaving the left alone gives

_____

Pushing electrons in the left ring and leaving the right alone gives

_____   _____

supply arrows

Note that one cannot push electrons in adjacent fused rings at the same time because one pair of pi electrons is held in common between the two.

41. Phenanthrene contains three fused rings and has five acceptable resonance structures. They can be generated as follows.

(a) Push electrons in the left ring.

_____   _____

supply arrows

(b) Push electrons in the right ring.

_____   _____

supply arrows

(c) Push electrons in the right and left rings.

_____   _____
         supply arrows

(d) Push electrons in the central ring.

_____   _____
         supply arrows

*   *   *   *   *

Writing resonance structures for fused aromatic compounds is particularly rich in opportunities for incorrect structures. After all resonance structures for fused aromatic compounds are written down, check for and remove the following.

1. Any structures with ten electrons around carbon, e.g.,

2. Any structures that do not have a continuous system of alternating single and double bonds, e.g.,

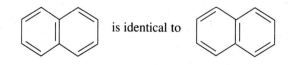

3. Any structures that are identical with others in the set, e.g.,

is identical to

Exercises

In these exercises, one Lewis structure of an aromatic compound is given along with the total number (including the one provided) of possible structures. Write the other structures.

42. Ethylbenzene (2)

43. p-xylene (2)

44. Anthracene (4)

More Complicated Ions

45. Very often more than two resonance structures can be written for an ion. One Lewis structure for the phenoxide ion is

(8)

The structure contains a pair of pushable electrons, namely, the unshared electrons on the negatively charged oxygen atom. The structure also contains a receptor, C_1, from which a pair of pi electrons can be pushed to C_2. This is shown as

 which generates

(9)

Structure (9) in its turn can be used to generate another resonance structure by pushing the unshared electrons at C_2 toward C_3 as the C_3–C_4 pi electrons are pushed to C_4.

generates

—————————————   —————————————
  supply arrows   **(10)**

A final structure, equivalent to but not identical to (9) can be arrived at by pushing the unshared pair at C_4 toward C_5 and moving the C_5–C_6 pi electrons to C_6.

generates

—————————————   —————————————
  supply arrows   **(11)**

Try to complete the following without referring to the structures you have just written. The complete resonance method designation for phenoxide ion, including both Kekulé forms of the first structure, is

46. One Lewis structure for the cyclopentadienide anion is

(12)

Four additional resonance structures can be written. The pair of unshared electrons at C_1 is pushed toward C_2, and the pair of pi electrons between C_2 and C_3 is pushed toward C_3.

(12)   **(13)**

In structure (13) there is a pair of electrons at C_3 and double bonds at C_4–C_5 and C_1–C_2. The unshared pair at C_3 is pushed toward C_4 and the pair of pi electrons at C_4–C_5 is pushed toward C_5.

supply arrows   **(14)**

(13)

This process can be repeated twice more to generate equivalent, but not identical, structures.

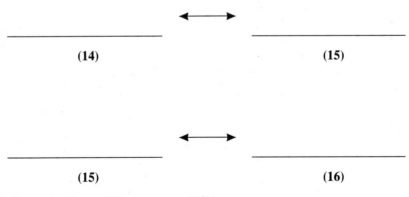

(14)   **(15)**

(15)   **(16)**

A similar operation on (16) regenerates (12).

47. There are five acceptable resonance structures for the benzyl cation.

(17)

First, the other Kekulé structure.

_____

(18)

Then, a pair of pi electrons from the ring can be pushed toward the positively charged carbon atom.

(17)   **(19)**

Structure (19) becomes (20) by pushing the C_3–C_4 pi electrons.

(19)   **(20)**

A similar operation on (20) generates (21), which is equivalent, but not identical, to (19).

(20)   **(21)**

48. As long as pushable electrons and receptors are available, one can continue to write new

resonance structures. For example, in the bromination of aniline the intermediate, (22), appears.

(22)

The pair of pi electrons between C_1 and C_2 can be pushed toward the positively charged C_3, which is a receptor.

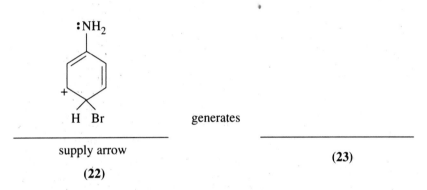

supply arrow

(22)   generates   **(23)**

In this new structure the same opportunity presents itself, i.e., to push the pair of pi electrons between C_5 and C_6 to the positively charged carbon at C_1.

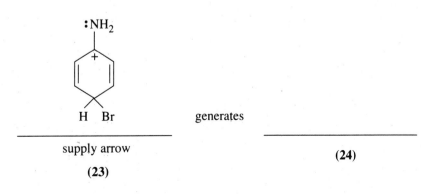

supply arrow

(23)   generates   **(24)**

In structure (23) there is another pair of pushable electrons next to the receptor, namely, the unshared pair on the nitrogen atom.

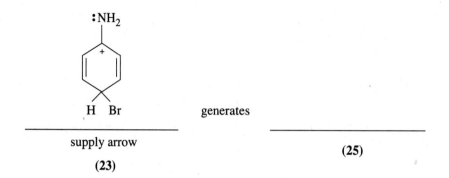

generates

_____

(25)

supply arrow

(23)

Don't refer back to complete the following. The complete resonance method notation for this intermediate is

_____   _____   _____   _____

49. The intermediate, (26), appears in the bromination of styrene.

(26)

The pair of electrons between C_1 and C_2 can be pushed toward C_3.

(26)   (27)

In (27) the same opportunity presents itself.

(27)   (28)

In (27) there is another pair of pushable electrons, namely, the pair of pi electrons in the double bond outside the benzene ring.

(27)   (29)

Exercises

One Lewis structure is provided along with the number of possible structures (including the one provided). Write the other structures.

50. (3)

51. (3)

52. (7)

53. (3)

55. (4)

54. (3)

Molecules Having Resonance Structures with Charge Separation

So far in this chapter all the resonance structures for a given molecule, ion, or radical have possessed the same formal charge distribution. These resonance structures are significant because the energies of the various structures are similar and, thus, each structure makes an important contribution to the hybrid.

In many cases, however, molecules for which there is a Lewis structure with no formal charge separation also have resonance structures with charge separation. These structures make smaller, albeit finite, contributions to the hybrid.

The Lewis structure for formaldehyde is

Of the two atoms, carbon and oxygen, that are sharing the pi electrons of the carbonyl bond, oxygen is the more electronegative. Therefore, it can act as a receptor. Pushing the pi electrons to the oxygen atom generates a new resonance structure.

The introduction of charge separation into a resonance structure usually increases the energy of that structure relative to one in which no separation of charge appears. The uncharged structure makes a substantially greater contribution to the hybrid. However, proposing that the charged structure makes a smaller but finite contribution emphasizes the **electrophilic** nature of a carbonyl carbon atom and the **nucleophilic** nature of the carbonyl oxygen atom. An alternative structure in which the pi electrons are pushed toward the carbon atom, i.e.,

No good!

is not considered because it makes no sense to push electrons away from the more electronegative atom.

Neither is it acceptable to propose a diradical resonance structure since that violates the principle that all resonance structures for a single molecule must contain the same number of paired and unpaired electrons.

No good!

56. The Lewis structure for acrolein (propenal) is

Pushing the pi electrons of the carbonyl bond toward the oxygen atom will generate a second resonance structure.

generates

_____   _____
        supply arrow

The second structure has a pair of pushable electrons (namely, the pi electrons of the carbon–carbon double bond) next to a receptor, the carbonyl carbon atom, which possesses a formal positive charge.

gives

_____   _____
        supply arrow

Don't refer back to complete the following. The resonance method notation for acrolein is

57. The Lewis structure for N-methylacetamide is shown. Pushing the pi electrons of the carbonyl bond toward the oxygen atom will generate a new resonance structure.

gives

_____
supply arrow

_____

The unshared electrons on the nitrogen atom in the second structure can be pushed to the positively charged carbonyl carbon and

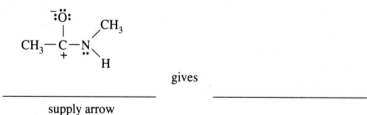

gives

_____
supply arrow

_____

Don't refer back to complete the following. The resonance method notation for N-methylacetamide is

58. Once again the presence of an aromatic ring greatly enhances the opportunities to push electrons and generate new resonance structures. The nitro group in nitrobenzene is a powerful **electron attractor**. A pair of pi electrons from the ring can be pushed toward the nitrogen atom as the pair

of pi electrons in the N–O double bond are pushed toward the oxygen atom. That is,

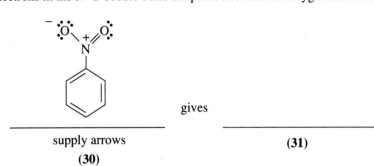

gives

———————————————————   ———————————————————
         supply arrows   **(31)**
         (30)

In (31) a pair of pi electrons in the ring can be pushed toward the positive charge.

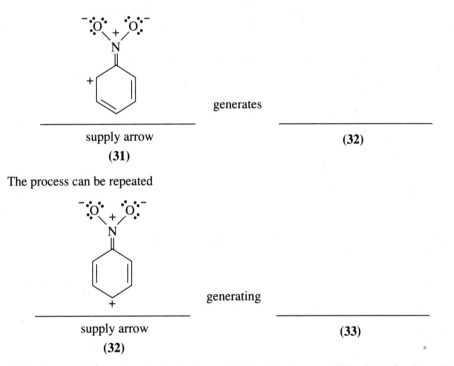

generates

———————————————————   ———————————————————
         supply arrow   **(32)**
         (31)

The process can be repeated

generating

———————————————————   ———————————————————
         supply arrow   **(33)**
         (32)

Fill in the complete resonance method notation for nitrobenzene. [Watch out for three other structures that can be generated from (30).]

———————————   ↔   ———————————   ↔   ———————————   ↔   ———————————

$\longleftrightarrow$ _____ $\longleftrightarrow$ _____ $\longleftrightarrow$ _____

59. Hypervalence (p. 12) results in an interesting resonance structure with separation of charge. A phosphonium ylide is an intermediate in the Wittig synthesis of alkenes. The phosphorous atom in the ylide is hypervalent. By pushing pi electrons away from the phosphorous one generates resonance structure with charge separation but with every atom obeying the octet rule.

$$(Ph)_3P = \overset{\curvearrowleft}{\underset{\underset{H}{|}}{C}} - H \longleftrightarrow \underline{\hspace{2cm}}$$

phosphonium ylide

Exercises

One Lewis structure is provided along with the total number of resonance structures (including the one provided). Write the others.

60. 2-cyclohexenone (3)

61. Ethyl acetate (3)

$$CH_3 - \overset{\overset{\displaystyle :\ddot{O}}{\|}}{C} - \ddot{\overset{..}{O}} - CH_2 - CH_3$$

62. Propionic acid (3)

$$H - \ddot{\overset{..}{O}} - \overset{\overset{\displaystyle \ddot{O}}{\|}}{C} - CH_2 - CH_3$$

63. Benzonitrile (7)

$$- C \equiv N\textbf{:}$$

64. An ylide (2)

$$(Ph)_3P = C \overset{\displaystyle H}{\underset{\displaystyle CH_3}{<}}$$

Answers

Chapter 2

1. CH₃—CH=CH—CH₂⁺ , CH₃—CH—CH=CH₂

2.

3. pi, oxygen,

    CH₃
       ＼
        C—Ö—H
       ╱   ⁺
    CH₃

4.

    ⁺Ö—H
       ‖
    CH₃—C—CH₂—CH₃ ,

    :Ö—H
       |
    CH₃—C—CH₂—CH₃
       ⁺

5.

6. oxygen, carbon, receptor,

    CH₃—Ö—CH₂⁺ , CH₃—Ö=CH₂⁺

7. carbon, CH₃—C≡Ö⁺

8.

    CH₃—CH—C=Ö⁺ , CH₃—CH—C≡Ö⁺
         | |
        CH₃   CH₃

9. pushable electrons, receptor

10.

11.

12. no additional structures

13.

14. no additional structures

15.

16.

17.

18.

19.

20.

21.

22.

23.

24.

25. no additional structures

26. no additional structures

27.

28. no additional structures

29. $CH_3-\overset{\bullet}{C}H-CH=CH_2$

30.

31.

32.

33.

34.

35.

36.

37.

38.

39.

40.

41.

42.   CH₂—CH₃

43.   CH₃

CH₃

44.

45.

(9)   (10)

(10)   (11)

46.

(13)   (13)   (14)

(14)   **(15)**   **(15)**   **(16)**

47.

(18)   **(19)**

          (20)

(19)

(20)   **(21)**

48.

(22)   **(23)**   **(23)**   **(24)**

(23)   **(25)**,

structures (22), (23), (24), and (25)

49.

(26)   **(27)**

(27)   **(28)**

(27)   **(29)**

50.

51.

A fourth structure,

is not included because the sp-hybridized nitrogen requires that it and the two adjacent carbon atoms be linear. This adds excessive ring strain to the structure.

52.

53.

54.

55.

56.

57.

58.

(30)   (31)

(31)   (32)

(32)   (33)

Structures (**30**) through (**33**) plus

59.

60.

61.

62.

63.

64. (Ph)₃P⁺—C̈—H
        |
        CH₃

Mechanisms

3

A mechanism is a step-by-step description of how a chemical reaction occurs. Each step will involve some kind of bond breaking and/or bond making. Pushing electrons is an excellent, graphic way to describe each step.

Development of experimental methods to elucidate mechanisms is one of the triumphs of science and was led by organic chemists. Thanks to the creativity and skill exercised by physical organic chemists over the last six decades, students can better understand how products arise from starting materials. Further, students can organize an otherwise intimidating amount of factual content into categories of mechanisms making the study of organic chemistry much more manageable. Chemists use mechanisms to predict new reactions as well as to understand old reactions.

In Chapter 1 we briefly trespassed into mechanisms when we pushed electrons to make ions and free radicals. In Chapter 2 you learned to push electrons in order to generate resonance structures. Now it is time to become an accomplished electron pusher. You can never be too skilled in pushing electrons.

Sigma Bond Breaking

Heterolytic cleavage of **sigma bonds** occurs under a variety of conditions. It is shown below. The arrow indicates that the sigma electrons that form the A–B bond are leaving A and becoming the exclusive property of B. Since the fragment A is formally losing one electron, it must become positively charged and B must become negatively charged since it gains an electron.

$$A-\overset{..}{\underset{..}{B}}: \longrightarrow A^+ \quad + \quad :\overset{..}{\underset{..}{B}}:^-$$

This occurs in the first step of the **familiar S$_N$1 reaction.** In *t*-butyl chloride, the leaving group is *chloride ion.* The *carbon–chlorine* bond cleaves, the electrons going with *chlorine,* which obtains a *negative* charge. The carbon atom obtains a *positive* charge. Thus, the cleavage of *t*-butyl chloride yields chloride ion plus the *t*-butyl cation.

1. In isopropyl chloride the leaving group is _____ ion. The _____ – _____ bond cleaves, the electrons going with _____, that obtains a _____ charge. The carbon atom obtains a _____ charge. Thus, the cleavage of isopropyl chloride yields chloride ion plus the _____ cation.

2. In cyclohexyl bromide the leaving group is _____ ion. The _____ – _____ bond cleaves, the electrons going with _____, that obtains a _____ charge. The carbon atom obtains a _____ charge. Thus, the cleavage of cyclohexyl bromide yields _____ ion plus the _____ cation.

_____   _____

supply arrow

*   *   *   *   *

Very often, heterolytic cleavage occurs from a charged intermediate which was formed in a previous step. The process is the same, but the charge on the products is different.

$$A - \overset{+}{\underset{..}{B}} - H \longrightarrow A^+ \quad + \quad :\overset{..}{\underset{..}{B}} - H$$

3. This kind of cleavage occurs in the second step of the dehydration of alcohols. In the first step the alcohol is protonated to form (1). In (1), the leaving group is a *water* molecule. The *carbon–oxygen* bond cleaves, the electrons going with the oxygen atom. The *oxygen* atom, which had

a *positive* charge, becomes *neutral*. The carbon atom obtains a *positive* charge. Thus, the cleavage of (1) yields *water* plus the *t-butyl* cation.

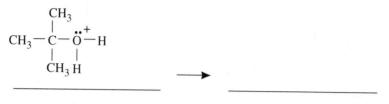

supply arrow

(1)

4. The acid-catalyzed hydrolysis of methoxymethyl acetate begins with protonation giving (2). A C–O bond cleaves, the electrons going with the _____ atom. The oxygen atom that had a positive charge becomes _____. The leaving group is *acetic acid*. The carbon atom from which acetic acid departed obtains a _____ charge. Thus, the cleavage of (2) yields acetic acid plus the *methoxymethyl* cation.

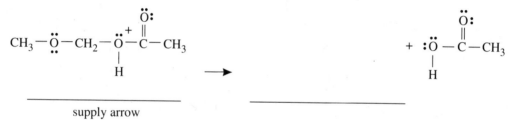

supply arrow

(2)

5. Sometimes a ring will be opened as a result of sigma bond breaking. The **protonated** epoxide, (3), will suffer heterolytic cleavage of one of its carbon–oxygen bonds, with the electrons going to the _____ atom. The oxygen atom, which had a _____ charge, becomes _____. The leaving group is

$$\diagdown \overset{\cdot\cdot}{\underset{\cdot\cdot}{O}} - H$$

The carbon atom from which the oxygen departed obtains a _____ charge.

supply arrow

(3)

Exercises

Here are some exercises in sigma bond breaking. Supply the arrows and the products. Remember that charge must be conserved. That is, if a neutral molecule is dissociated, the algebraic sum of the charges on the products must equal zero. If a positively charged ion is dissociated, the algebraic sum of the charges on the products must equal +1.

6.

7.

8.

9.

10.

11.

12.

13.

14.

Sigma Bond Making

Formation of a sigma bond occurs when an anion and a cation encounter one another. That is

$$A^+ \quad :\ddot{B}:^- \quad \longrightarrow \quad A-\ddot{B}:$$

The arrow indicates that a pair of electrons which was the exclusive property of B is now shared by A and B.

The product-forming step in many S_N1 reactions involves the formation of a sigma bond between the intermediate carbocation and some negatively charged nucleophile. For example, α-phenylethyl cation will yield an ester by reaction with acetate ion. A pair of unshared electrons on the negatively charged *oxygen* atom forms a bond between *oxygen* and *carbon* and becomes a shared pair of sigma electrons.

The oxygen atom that had a *negative* charge becomes *neutral*, and the carbon atom that had a *positive* charge becomes *neutral*.

15. *Sec*-butyl cation will react with bromide ion to form *sec*-butyl bromide. A pair of _____ electrons on the negatively charged _____ ion forms a bond between _____ and _____ and becomes a pair of shared sigma electrons. The bromine atom that had a _____ charge becomes _____.

_____
              supply arrow

     *   *   *   *   *

More commonly, sigma bonds are formed by the reaction of carbocations with neutral molecules possessing unshared pairs. The result is a positively charged intermediate.

16. During the solvolysis of alkyl halides in ethanol, a carbocation forms a sigma bond with one of the unshared pairs on the alcohol oxygen atom. A pair of unshared electrons on the oxygen atom forms a bond between *oxygen* and *carbon* and becomes a shared pair of *sigma* electrons. The oxygen atom, which was *neutral,* now possesses a formal *positive* charge. The carbon atom, which had a *positive* charge, is now *neutral.*

_____
              supply arrow

17. Carbocations react with amines in an analogous process. The _____ electrons on the nitrogen atom form a bond between _____ and _____ and becomes a pair of _____ electrons. The nitrogen atom, which was neutral, now possesses a formal _____ charge. The carbon atom which had a _____ charge is now _____.

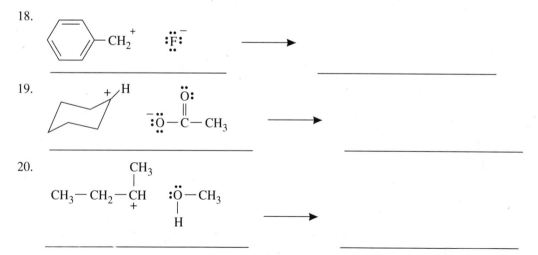

supply arrow

*　　*　　*　　*　　*

Note some important features of the way electron pushing is used to write mechanisms.

1. **Electrons are always pushed *toward* a center of positive charge.**

2. **Although the position of positive charge changes, the positive charge is *not* pushed. Arrows should never lead away from a positive charge.**

3. **Total charge is conserved. That is, the sum of the charges on one side of the equation always equals the sum of the charges on the other.**

4. **Electrons are conserved. That is, the number of valence electrons on each side of an equation is always equal.**

Exercises

Here are some examples of sigma bond making. Supply the arrows and the products.

18.

$-CH_2{}^+$　　$:\ddot{F}:^-$　　⟶

_____　　　　_____

19.

$+$ ^{H}　　　$\overset{\overset{\ddot{O}:}{\|}}{:\ddot{O}-C-CH_3}$　　⟶

_____　　　　_____

20.

$$CH_3-CH_2-\underset{+}{\overset{\overset{\displaystyle CH_3}{|}}{CH}} \quad \underset{|}{\overset{}{:\ddot{O}-CH_3}}$$
$$H$$

_____　　　　_____

21.

_____   _____

22.

$$CH_2 - CH_2^+$$
$$|$$
$$H - \underset{..}{\overset{..}{O}}:$$

⟶

_____   _____

Simultaneous Bond Making and Breaking

While isolated events of bond breaking and bond making are important in writing mechanisms of organic reactions, it is far more common to encounter steps in which bond making and bond breaking occur simultaneously. In some instances, bond making has proceeded well along toward completion and bond breaking has only begun as the transition state is reached. The opposite also occurs. These are fine points of mechanism and need not concern a student just learning to push electrons.

Sigma Bond Making and Breaking

The simultaneous making and breaking of sigma bonds is illustrated nicely in the familiar S_N2 re-action. A negatively charged nucleophile, Nu^-, approaches a carbon atom having a leaving group, $-L$, in a direction anti to and rearward to the leaving group. The pushable electrons are possessed by the nucleophile and the receptor is the carbon atom that itself has a pair of pushable electrons, namely, the sigma electrons forming the C–L bond. The products are a compound having a C–Nu bond and the anion of the leaving group. Thus, Nu has been substituted for L.

This illustrative equation has been written in the **stereochemically** accurate form showing that an inversion of configuration occurs during S_N2 reactions. It was done this way to make the illustration as clear as possible. However, the student should note that the S_N2 reaction is usually written without reference to stereochemistry unless a stereochemical point is being made. Thus, the reaction is usually written

The S_N2 reaction of methyl iodide with bromide ion illustrates the process. The nucleophile is _bromide ion_. The leaving group is _iodide ion_. A pair of unshared electrons on the bromide ion is pushed

toward the carbon atom and, simultaneously, the **sigma electrons** of the carbon–iodine bond are pushed toward *iodine*. The result is the making of a sigma bond between bromine and carbon, and the breaking of a sigma bond between carbon and iodine. The bromine, which had a *negative* charge, becomes *neutral* and the iodine, which was *neutral,* now possesses a formal *negative* charge.

23. Another example is the reaction of hydroxide ion with ethyl chloride. The nucleophile is _____ ion. The leaving group is _____ ion. A pair of _____ electrons from hydroxide ion is pushed toward the carbon atom and, simultaneously, the _____ electrons of the carbon–chlorine bond are pushed toward _____. The result is the making of a sigma bond between oxygen and _____ and the breaking of the sigma bond between carbon and _____. The oxygen, which had a negative charge, is now _____ and the chlorine, which was neutral, now possesses a formal _____ charge.

$$H-\ddot{O}:^{-} \qquad \begin{matrix} CH_2-\ddot{C}l: \\ | \\ CH_3 \end{matrix} \qquad \longrightarrow$$

_____   _____

          supply arrows

24. Consider the reaction of hydroxide ion with benzyl bromide.

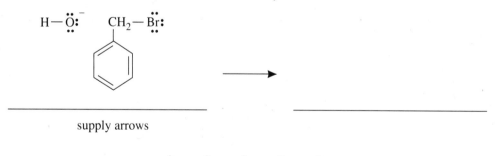

_____   _____

          supply arrows

*   *   *   *   *

Carbon need not always be the center of this type of reaction. The removal of an acidic proton from an organic compound by a base is a common occurrence.

25. The reaction of phenol with hydroxide ion results in the phenoxide ion plus a molecule of water.

A pair of unshared electrons on the oxygen atom of the hydroxide ion is pushed toward the hydrogen atom and, simultaneously, the *sigma* electrons of the bond between the hydrogen and oxygen atoms of phenol are pushed toward *oxygen.* The result is the making of a sigma bond between oxygen and _____ and the breaking of a sigma bond between hydrogen and _____. The oxygen atom of the hydroxide ion, which had a *negative* charge, is now *neutral,* and the oxygen atom of phenol, which was *neutral,* now possesses a formal *negative* charge.

26. The amide ion is a strong enough base to remove a proton from acetylene. The reaction results in a molecule of ammonia plus the acetylide anion. A pair of _____ electrons on the _____ atom of the amide ion is pushed toward the hydrogen atom, and, simultaneously, the _____ electrons of the carbon–hydrogen bond of acetylene are pushed toward _____. Thus, a sigma bond is made between nitrogen and _____, and a sigma bond is broken between hydrogen and _____. The nitrogen atom that possessed a negative charge in the amide ion is now _____, and the carbon atom that was neutral in acetylene now possesses a _____ charge.

_____   _____

supply arrows

27. The reaction of ethoxide ion with one of the alpha hydrogens of acetaldehyde is similar. The products are a molecule of ethanol and the anion of acetaldehyde.

_____   _____

supply arrows

*   *   *   *   *

The formation of a carbon–carbon sigma bond occurs in the reaction of a Grignard reagent with ethylene oxide. In the illustration the **Grignard reagent,** ethyl magnesium bromide, is written as an ion pair consisting of the MgBr⁺ ion and the ethyl anion. The nucleophile is the *ethyl anion.* A pair of unshared electrons on the carbanion is pushed toward one of the *carbon* atoms of ethylene oxide. Simultaneously, the *sigma* electrons of the carbon–oxygen bond are pushed toward *oxygen*. The result is a ring opening. The carbanion, which had a *negative* charge, has become *neutral* and the oxygen atom of ethylene oxide, which was *neutral*, now possesses a formal *negative* charge.

28. In the reaction of propyl magnesium iodide with ethylene oxide, the nucleophile is the _____ anion. A pair of _____ electrons on the carbanion is pushed toward one of the _____ atoms of ethylene oxide. Simultaneously, the _____ electrons of the carbon–_____ bond are pushed toward _____. Thus, a sigma bond has been made between carbon and _____, and a sigma bond has been broken between carbon and _____. The carbanion, which had a negative charge, has become _____, and the oxygen atom of ethylene oxide, which was neutral, now possesses a _____ charge.

$$CH_3-CH_2-\overset{-\,\cdot\cdot}{C}H_2 \quad \underset{CH_2}{\overset{CH_2}{\diagup\diagdown}}\ddot{O}\colon \longrightarrow$$
$$\overset{+}{MgI}$$

_____   _____

        supply arrows

29. Analogously, an alcoholate anion will open an epoxide ring. In the reaction of ethoxide ion with 2-butene oxide, the nucleophile is _____ ion. A pair of _____ electrons on the _____ atom of the ethoxide ion is pushed toward one of the _____ atoms in 2-butene oxide. Simultaneously, the _____ electrons of the carbon–_____ bond are pushed toward _____. Thus, a sigma bond has been made between oxygen and _____, and a sigma bond has been broken between carbon and _____. The ethoxide ion, which had a negative charge, has become _____, and the oxygen atom of 2-butene oxide, which was neutral, now possesses a _____ charge.

$$CH_3-CH_2-\overset{-\,\cdot\cdot}{\underset{\cdot\cdot}{O}}\colon \quad \begin{array}{c} CH_3 \\ | \\ CH \\ | \quad \diagdown \\ CH \quad \diagup O\colon \\ | \\ CH_3 \end{array} \longrightarrow$$

_____   _____

        supply arrows

The simultaneous making and breaking of sigma bonds occurs also between a neutral nucleophile or base with pushable electrons and a positively charged species that can act as a receptor.

$$\ddot{N}u \qquad A\!-\!X^{+} \longrightarrow \overset{+}{N}u\!-\!A \quad + \ddot{X}$$

In organic chemistry the most common examples of this are reactions in which a molecule possessing unshared electrons is protonated by a positively charged acid.

Alcohols are protonated by hydronium ion (H_3O^+). In the reaction of *t*-butyl alcohol with hydronium ion the base is *t*-butyl alcohol, which has unshared pairs on the *oxygen* atom. The acid is *hydronium ion.* The unshared electrons on the oxygen atom of *t*-butyl alcohol are pushed toward the *hydrogen* atom. Simultaneously, the pair of *sigma* electrons between the *hydrogen* and *oxygen* atoms of the hydronium ion is pushed toward the *oxygen* atom. Thus, an *oxygen–hydrogen* sigma bond is made and a *hydrogen–oxygen* sigma bond is broken. The oxygen atom of the alcohol, which was *neutral,* now possesses a formal *positive* charge and the oxygen atom of the hydronium ion, which had a *positive* charge, is now *neutral.*

30. Cyclohexanol can be protonated by hydronium ion. The base is _____, which has unshared pairs of electrons on the _____ atom. The acid is _____ ion. A pair of _____ electrons on the oxygen atom of cyclohexanol is pushed toward the _____ atom. Simultaneously, the pair of _____ electrons between the hydrogen and _____ atoms of the hydronium ion is pushed toward the _____ atom. Thus, an oxygen–_____ sigma bond is made, and a hydrogen–_____ sigma bond is broken. The oxygen atom of the alcohol, which was neutral, now possesses a _____ charge, and the oxygen atom of the former hydronium ion, which had a positive charge, is now _____.

_____   _____

      supply arrows

31. The reaction just described is reversible. Deprotonation of the conjugate acid of an organic base

by water provides another example of simultaneous making and breaking of sigma bonds. Thus, in the deprotonation of anilinium ion by water, the base is water, which has unshared electrons on the _____ atom. The acid is _____ ion. A pair of _____ electrons on the oxygen atom of water is pushed toward the _____ atom. Simultaneously, the pair of _____ electrons between the hydrogen and _____ atoms of the anilinium ion is pushed toward the _____ atom. Thus, the oxygen–_____ sigma bond is made and a hydrogen–_____ sigma bond is broken. The nitrogen atom, which possessed a positive charge, is now _____, and the oxygen atom, which was neutral, now possesses a formal _____ charge.

_____   _____

           supply arrows

32. Let us consider an entire mechanism of an organic reaction. The acid-catalyzed ring opening of epoxides is initiated by protonation on the epoxide oxygen by hydronium ion.

_____   _____

        supply arrows

33. The resulting ion then suffers nucleophilic attack on carbon by the neutral molecule, water, and the ring is opened.

_____   _____

        supply arrows

34. The resulting ion is then deprotonated by water giving a hydronium ion plus the organic product, a neutral 1,2-diol.

_____   _____

                 supply arrows

Thus, all three steps in this organic reaction are examples of simultaneous sigma bond making and sigma bond breaking.

Exercises

Here are some examples of simultaneous sigma bond making and sigma bond breaking. Where arrows are supplied, write in the products. Where the products are written in, supply the appropriate arrows.

39.

40.

41.

42.

Homolytic Sigma Bond Making and Breaking

We first saw the production of free radicals in Chapter 1. Each case involved the homolysis of a molecule in which all the electrons were paired. Uncoupling a pair of electrons requires fairly high energy. The examples in Chapter 1 used elevated temperatures or ultraviolet light. The best examples involved the scission of weak covalent bonds.

Free radical reactions occupy an important place in organic chemistry and are not so rare as the previous paragraph might suggest. Once the first few radicals are produced *(initiation),* they eagerly react with covalent bonds to produce more radicals *(propagation).* A sigma bond is broken and another is made.

A bromine radical from the photochemical homolysis of Br_2 reacts with one of the carbon–hydrogen bonds of methane.

A hydrogen–bromine sigma bond is formed and a methyl radical remains.

43. A chlorine radical and ethane form HCl and an ethyl radical.

_____ + _____

44. A chlorine radical and cyclohexane yield the cyclohexyl radical.

    + H—$\overset{\cdot\cdot}{\underset{\cdot\cdot}{Cl}}$:

_____   _____

supply arrows

45. A methyl radical and a bromine molecule give bromomethane and a bromine radical.

          +

_____   _____

46. The hydrogen–silicon bond is susceptible to homolysis.

$$Cl_3Si—H \quad \cdot\overset{\cdot\cdot}{\underset{\cdot\cdot}{O}}—\overset{\overset{\displaystyle \overset{\cdot\cdot}{O}:}{\|}}{C}—Ph \longrightarrow \quad + \quad H—\overset{\cdot\cdot}{\underset{\cdot\cdot}{O}}—\overset{\overset{\displaystyle \overset{\cdot\cdot}{O}:}{\|}}{C}—Ph$$

_____   _____

supply arrows

47. Cleavage of the hydrogen–tin bond yields tin radicals.

$$CH_3—\overset{\overset{\displaystyle CN}{|}}{\underset{\underset{\displaystyle CH_3}{|}}{C}}\cdot \quad H—Sn—R_3$$

⟶   +

_____   _____   _____

supply arrows

Exercises

Supply the arrows and the products for each of these reactions as described.

48. A chlorine radical removes a hydrogen from propane to form the isopropyl radical.

H— C̈l: +

_____   _____

49. A chlorine radical removes a hydrogen from propene to form the allyl radical.

:C̈l·   H—CH$_2$—CH=CH$_2$ ⟶   :C̈l—H +

_____   _____

50. A benzyl radical is formed from toluene.

+   H— B̈r:

_____   _____

51. Bromine is removed homolytically from 6-bromo-1-hexene.

CH$_2$=CH—(CH$_2$)$_3$—CH$_2$— B̈r: · SnR$_3$ ⟶   +

_____   _____   _____

52. A radical center is formed on an allylic position of cyclopentene.

+

_____   _____   _____

53. Iodine is removed homolytically from iodocyclohexane.

   · SnR$_3$ ⟶   +

_____   _____   _____

54. The carbon radical formed in 52 removes bromine from N-bromosuccinimide.

_____   _____ _____

Sigma Bond Making and Pi Bond Breaking

There are many steps in organic reactions in which a sigma bond is formed at the same time that a pi bond is broken. In each case, this results in a condensation of a molecule and an ion into a single ion. Actually, this category of reactions can be divided further according to whether the electrons being pushed are unshared electrons or pi electrons.

A negatively charged nucleophile may react with a pi bond as follows.

In this case, the electrons being pushed are the unshared electrons on the nucleophile. The receptor is the carbon atom in the double bond, which has a pair of pushable electrons itself. Pushing the unshared electrons to the carbon atom results in the Nu–C sigma bond. Pushing the pi electrons away from the carbon atom breaks the pi bond and places a formal negative charge on the atom X. Overall, the single negative charge has been conserved.

55. The first step in the saponification of ethyl acetate involves a reaction in which hydroxide ion attacks the carbonyl-carbon atom. The nucleophile is a _____ ion. A pair of _____ electrons on the oxygen atom of hydroxide ion is pushed toward the carbonyl–carbon atom. The pair of *pi* electrons of the carbon–oxygen double bond of ethyl acetate is pushed toward the _____ atom. A sigma bond has been made between oxygen and *carbon*. A pi bond has been broken between carbon and _____. The hydroxyl oxygen, which had a negative charge, is now *neutral,* and the carbonyl oxygen, which was *neutral,* now possesses a _____ charge.

56. Similarly, amides are attacked by hydroxide ion. The nucleophile is _____ ion. A pair of _____ electrons on the hydroxyl oxygen atom is pushed toward the carbonyl-carbon atom. The pair of _____ electrons in the carbon–_____ double bond is pushed toward the _____ atom. A sigma bond has been made between oxygen and _____. A pi bond has been broken between carbon and _____. The hydroxyl oxygen, which had a negative charge, is now _____. The carbonyl oxygen, which was neutral, now possesses a _____ charge.

_____   _____
           supply arrows

57. The first step in the formation of a cyanohydrin under neutral or basic conditions involves a similar process.

_____   _____
           supply arrows

58. Since the alkyl portion of a Grignard reagent can be treated as a carbanion, its well known reaction with a carbonyl group fits this category.

_____   _____
           supply arrows

59. The pi bond that is broken can be part of a triple bond. Consider the first step in the base-catalyzed hydrolysis of a nitrile.

_____   _____

supply arrows

*   *   *   *   *

A slight variation of this process occurs when the nucleophile possessing the unshared electrons is neutral. In virtually all of these cases, the pi systems must have suffered some kind of previous activation in order for the reaction to proceed at a reasonable rate. The activation process (usually protonation) puts a positive charge on the pi system.

60. We have already seen that hydroxide ion, a strong nucleophile, will attack the carbonyl – carbon atom of an ester. Analogously, water, a weak nucleophile, will attack the carbonyl–carbon atom of a protonated ester. A pair of unshared electrons on the oxygen atom of the water molecule is pushed toward the carbonyl–carbon atom. The pair of *pi* electrons of the carbon–oxygen double bond is pushed toward the *oxygen* atom. Thus, a sigma bond has been made between oxygen and _____, and a pi bond has been broken between carbon and _____. The oxygen atom of the water molecule, which was neutral, now possesses a _____ charge. The carbonyl–oxygen atom, which had a positive charge, is now _____.

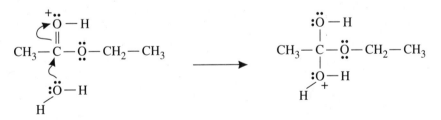

61. An important step in the production of acetals occurs when a molecule of alcohol undergoes reaction with the protonated carbonyl group of an aldehyde. A pair of _____ electrons on the _____ atom of the alcohol is pushed toward the carbonyl–carbon atom. The pair of _____ electrons of the carbon–oxygen double bond is pushed toward the _____ atom. Thus, a sigma bond is made between oxygen and _____, and a pi bond has been broken between carbon and _____.

The oxygen atom of the alcohol, which was neutral, now possesses a _____ charge. The oxygen atom of the protonated aldehyde, which had a positive charge, is now

_____.

_____   _____

        supply arrows

62. The acid-catalyzed hydrolysis of nitriles involves this kind of reaction on a triple bond.

_____   _____

        supply arrows

           *   *   *   *   *

We have seen a number of examples of the simultaneous sigma bond making and pi bond breaking type of reaction where the pushable electrons were unshared electrons. In many mechanisms, one encounters steps in which there is simultaneous sigma bond making and pi bond breaking and it is the pi electrons that are pushed.

Once again, an ion and a molecule condense into a single ion. The electrons being pushed are the pi electrons of an unsaturated system. The receptor is an electrophile, E, usually possessing a formal positive charge. Pushing the pi electrons to the electrophile results in the E–C sigma bond and breaks the pi bond. A formal positive charge is left on the atom Z, thus conserving the charge in the system.

In a typical Friedel–Crafts alkylation of benzene, a *t*-butyl cation reacts with a molecule of benzene. A pair of *pi* electrons from the aromatic ring is pushed toward the positively charged *carbon* atom of the *t*-butyl cation. Thus, a pi bond between carbon and *carbon* is broken, and a sigma bond between carbon and *carbon* is made. The positively charged carbon atom of the *t*-butyl cation is now *neutral*. The carbon atom, *from which the pi electrons departed*, now possesses a *positive* charge.

Here we see why it is necessary to preserve the Kekulé structure for benzene (Chapter 2). For clarity it is expedient to push electrons in pairs, and that is how it is done above. The reaction, as it actually occurs, is quite a complicated affair involving the entire pi system of the benzene ring.

63. Consider the alkylation of *p*-xylene by an isopropyl cation. A pair of _____ electrons from the aromatic ring is pushed toward the positively charged carbon atom of the isopropyl cation. Thus, a pi bond is broken between carbon and _____, and a sigma bond is made between carbon and _____. The positively charged carbon atom of the isopropyl cation is now _____. The carbon atom, from which the pi electrons departed, now possesses a _____ charge.

supply arrow

64. The condensation step in the dimerization of 2-methylpropene is very similar. The pi electrons of the double bond in a 2-methylpropene molecule are pushed toward the *t*-butyl cation. However, there are two ways in which this can be done. One way results in a tertiary cation.

supply arrow

Another way results in a primary cation.

supply arrow

The stabilities of the resulting carbocation (actually, the stabilities of the transition states) determine that the former process predominates.

65. It is convenient to view protonation of a pi system as occurring by this simple process. Thus, a pair of _____ electrons from 2-butene is pushed toward the proton. A carbon–_____ pi bond has been broken, and a carbon–_____ sigma bond has been made. The proton, which had a positive charge, is now _____. The carbon atom from which the pi electrons departed now possesses a _____ charge.

$$CH_3-CH=CH-CH_3$$

$$H^+$$

_____   _____

supply arrow

66. The protonation of 1-methylcyclohexene occurs in a similar fashion. Here you must make the proper choice between making a bond between the proton and C_1 or the proton and C_2.

_____   _____

supply arrow

However, as will be shown in a later section, protonation is usually more complicated since it is inaccurate to propose an uncombined proton as a viable species.

Pi Bond Making and Sigma Bond Breaking

There are many steps in organic reaction mechanisms where a pi bond is formed at the same time that a sigma bond is broken. Each of the reactions in the preceding section, when written in reverse, provides an example of this process.

The pushable electrons are the unshared electrons on the atom Z. As they are pushed toward the carbon atom to form a pi bond, the sigma electrons that form the Nu–C bond leave with Nu. Charge is conserved as Nu leaves as an anion.

Consider the fate of the ion formed from the attack of hydroxide ion on ethyl acetate. If one of the pairs of unshared electrons on the negatively charged oxygen atom is pushed toward carbon to reform the carbonyl pi bond and the sigma bond between carbon and the OH group is broken, we have exactly reversed the reaction by which the ion was formed.

This step occurs. However, since it does not lead toward the products of saponification, namely, carboxylate anion and alcohol, we shall consider another fate for the ion.

67. If, as the unshared electrons are being used to reform the carbonyl pi bond, the sigma bond between the carbon and the $-\ddot{\text{O}}-CH_2-CH_3$ group is broken, we get something that looks like progress toward products. Thus, a pi bond has been made between oxygen and _____, and a sigma bond has been broken between carbon and _____.

supply arrows

The simple transfer of a proton from the carboxylic acid to the alcoholate anion leads to the saponification products.

68. The final step in the benzoin condensation involves the making of a pi bond between oxygen and carbon and the breaking of the sigma bond between carbon and the CN group.

supply arrows

*   *   *   *   *

Another general form that these steps can take involves the expulsion of a neutral nucleophile.

69. The cleavage step in the acid-catalyzed hydrolysis of esters provides a good example. A pair of unshared electrons from one of the OH groups is pushed toward the *carbon* atom. Simultaneously, the pair of *sigma* electrons is pushed toward the positively charged oxygen atom. Thus, a pi bond is made between oxygen and _____, and a sigma bond is broken between carbon and _____. The neutral oxygen atom from which the pi electrons were pushed now possesses a _____ charge. The positively charged oxygen toward which the electrons were pushed is now _____.

Here, the resulting conjugate acid of acetic acid loses a proton to complete the hydrolytic mechanism.

70. Similarly, the penultimate step in the hydrolysis of an acetal is the loss of a methanol molecule.

_____   _____

supply arrows

*   *   *   *   *

The examples of simultaneous pi bond formation and sigma bond breaking that we have seen so far have occurred when a pair of unshared electrons was pushed to form the pi bonds, and a nucleophile, either neutral or negatively charged, was expelled. Another way in which a pi bond can be formed when a sigma bond is broken is to push a pair of sigma electrons in such a way as to form the pi bond. In these cases a positively charged fragment is lost.

$$E\!-\!C\!-\!Z^{+} \longrightarrow E^{+} \quad \diagdown C\!=\!Z$$

Most commonly, this occurs when a proton is lost from some positively charged intermediate.

71. The *t*-butyl cation can lose a proton from a carbon next to the carbocation and form the neutral molecule, isobutylene. Thus, a pair of sigma electrons forming a *carbon–hydrogen* bond is pushed toward the positively charged *carbon* atom. The sigma bond between hydrogen and _____ is broken, and a pi bond is formed between carbon and _____. The neutral hydrogen atom now possesses a _____ charge. The positively charged carbon atom is now _____.

72. Similarly, a proton can be lost from *sec*-butyl cation to form 2-butene. A pair of sigma electrons which form the hydrogen-carbon bond is pushed toward the positively charged _____ atom. The sigma bond between hydrogen and _____ is broken, and a pi bond between carbon and _____ is made. The neutral hydrogen atom now possesses a _____ charge. The positively charged carbon atom is now _____.

supply arrow

73. Consider the loss of a proton from the carbocation below. This intermediate could lose a primary hydrogen giving the terminal olefin, 2,4,4-trimethyl-1-pentene.

supply arrow

Conversely, it could lose a secondary hydrogen and become 2,4,4-trimethyl-2-pentene.

$$CH_3-\overset{CH_3}{\underset{\underset{+}{}}{C}}-\overset{}{\underset{\underset{H}{}}{CH}}-\overset{CH_3}{\underset{\underset{CH_3}{}}{C}}-CH_3 \longrightarrow$$

supply arrow

The latter process predominates since the olefin formed is more stable. The reformation of the aromatic system in the last step of electrophilic aromatic-substitution involves this process.

74. In the final step of the chlorination of benzene, the positively charged intermediate loses a proton from the carbon atom to which a chlorine atom has been attached. The pair of sigma electrons that form the bond between *carbon* and *hydrogen* is pushed toward the positively charged carbon atom. The result is the breaking of a hydrogen–_____ sigma bond and the making of a carbon–_____ pi bond. (The 6-pi electron aromatic system is restored.) The neutral hydrogen atom now possesses a _____ charge and the positively charged carbon atom is now _____.

75. The final step in the nitration of toluene is similar.

_____
              supply arrow

76. Also, the final step in the bromination of *m*-xylene.

_____
              supply arrow

77. On rare occasions a *t*-butyl group is lost in electrophilic aromatic substitution.

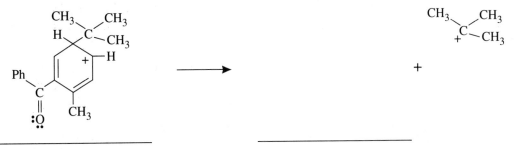

_____   _____
       supply arrow

78. Finally, the loss of a proton from the conjugate acid of a carbonyl group is often found as the final step in a reaction mechanism.

_____   _____
       supply arrow

79. Similarly,

_____   _____
       supply arrow

Exercises

Where arrows are supplied write in the products; where the products are written supply the appropriate arrows.

80.

_____

81.

_____

82.

_____

83.

_____

84.

_____

85.

_____

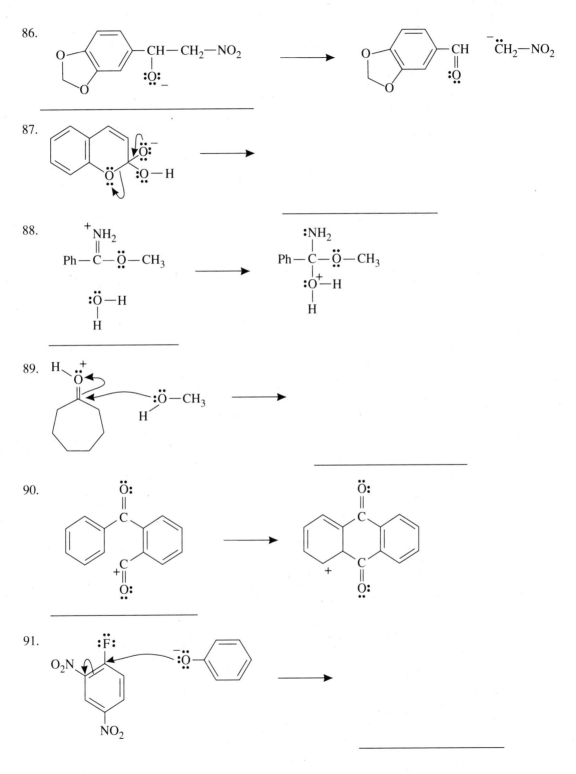

86.

87.

88.

89.

90.

91.

92.

Homolytic Sigma Bond Making and Pi Bond Breaking

Free radicals will react with pi systems. Such reactions usually result in addition to double bonds.

Hydrogen bromide will add to 2-butene in a free radical reaction. Initiation provides bromine radicals.

The bromine radical takes one of the pi electrons from 2-butene, leaving an unpaired electron on carbon.

The carbon radical abstracts a hydrogen from H–Br, giving the addition product, 2-bromobutane, plus another bromine radical that can propagate the reaction.

93. (a) Show a similar reaction for cyclohexene.

(b) Show the reaction of the carbon radical with H–Br to propagate the reaction.

supply arrows

When the alkene is unsymmetrical, the reaction takes one of two possible pathways.

94. (a) The radical attack might lead to the primary carbon radical.

_____
supply arrows

(b) Actually, attack leads to the more stable secondary radical.

CH_2=CH—CH_3

:Ḃr:

_____   _____
supply arrows

Other radicals add to double bonds.

95.

_____

96.

$Cl_3Si•$ CH_2=CH—$CH$$\overset{CH_3}{\underset{CH_3}{<}}$ ⟶

_____   _____
supply arrows

97. Carbon radicals will add to carbon–carbon double bonds.

⬠• CH_2=CH—CN ⟶

_____   _____
supply arrows

98. (a) A radical will attack carbon–sulfur double bonds.

_____   _____

          supply arrows

(b) Actually, part (a) is the first step in a clever way to produce R•. See if you can push electrons to produce the products.

_____

          supply arrows

Complex Mechanisms

When pushing electrons in this chapter, we have, until now, illustrated steps in reaction mechanisms with examples in which the making and breaking of no more than two bonds occurs in any step. The awful truth is that there are many examples that are more complex. We shall take up some of them in this section. It should be emphasized that these steps are more complicated in numbers of electron pairs being pushed, but the types of bond making and bond breaking should all be familiar by this time.

The E2 elimination is a reaction in which everything happens at once. In the example below a base abstracts a proton _beta_ to the leaving group, X. The electrons that had formed the C–H bond are pushed toward the _alpha_ carbon from which, in turn, the leaving group departs, taking with it the pair of electrons that had formed the C–X bond. All of this happens more or less simultaneously. Inspection reveals that this process involves sigma bond making, *sigma* bond breaking, pi bond making and sigma bond breaking.

99. In the formation of 2-butene from 2-bromobutane a pair of unshared electrons from the oxygen atom of hydroxide ion is pushed toward the beta hydrogen forming an oxygen–*hydrogen* bond. The pair of sigma electrons which was the hydrogen–carbon bond is pushed toward the alpha carbon atom forming a pi bond between carbon and *carbon*. The sigma electrons of the carbon–bromine bond are pushed toward the *bromine* atom. The hydroxyl oxygen, which had a negative charge, is now *neutral* and the bromine atom, which was neutral, now has a *negative* charge.

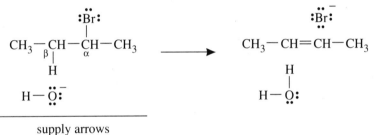

supply arrows

100. In the reaction of ethoxide ion with *t*-butyl chloride, a pair of unshared electrons on the oxygen atom of ethoxide ion is pushed toward the beta hydrogen forming an oxygen–_____ bond. The pair of sigma electrons that was the hydrogen–carbon bond is pushed toward the alpha carbon forming a pi bond between carbon and _____. The sigma electrons of the carbon–chlorine bond are pushed toward the chlorine atom. The ethoxyl oxygen, which had a negative charge, is now _____. The chlorine, which was neutral, now has a _____ charge.

supply arrows

101. Here is an example of an E2 reaction forming a double bond in a ring.

⟶

_____   _____
           supply arrows

102. The Hofmann elimination is a similar reaction except that the charge distribution is different. The hydroxyl oxygen, which had a negative charge, becomes _____, and the nitrogen, which had a positive charge, becomes _____.

$$\overset{+}{N}(CH_3)_3$$
$$CH_2-CH-CH_2-CH_3$$

H—Ö:⁻

⟶

_____   _____
           supply arrows

103. Here is a Hofmann elimination in which a double bond is formed in the ring.

⟶

_____   _____
           supply arrows

104. Finally, the Hofmann elimination can result in a ring-opening reaction.

_____   _____
             supply arrows

105. When a chlorine molecule interacts with a carbon–carbon double bond, the result is a **carbocation.** The pi electrons of the carbon–carbon double bond are mobilized to form a carbon–chlorine sigma bond. In turn, the sigma electrons of the chlorine–chlorine bond are pushed so that the chloride ion departs. This process involves pi bond breaking, sigma bond making, and sigma bond breaking.

_____
             supply arrows

The carbon atom, C_2, from which the pi electrons depart obtains a _____ charge, and the chlorine atom to which the electrons are pushed obtains a _____ charge.

106. A unionized acid protonates a carbon–carbon double bond in an analogous fashion.

_____
             supply arrows

107. Another example is

_____   _____
             supply arrows

*   *   *   *   *

When the carbon–carbon double bond is unsymmetrical, this type of reaction can lead to two different carbocation intermediates. The choice of which one is preferred is determined by the relative stabilities of the carbocations.

108. The protonation of propylene by hydrogen chloride can lead to two different carbocations. If as the pi bond breaks, the new sigma bond is made between C_1 and the hydrogen atom, a **secondary carbocation** is formed.

However, if the new sigma bond is made between C_2 and the hydrogen atom, a **primary carbocation** is formed.

Because the transition state leading to the secondary carbocation is more stable, the first of these two products is formed preferentially.

109. When hydrogen iodide is allowed to react with isobutylene, one of two different carbocations can form. If the new sigma bond is formed between C_1 and the hydrogen atom, a **tertiary carbocation** results.

    supply arrows

However, if the new sigma bond is formed between C$_2$ and the hydrogen atom, a primary carbocation is produced.

supply arrows

The first of these two is preferred.

110. The reaction of bromine with styrene can yield a secondary, benzylic carbocation.

supply arrows

or, a primary carbocation,

supply arrows

The first reaction is preferred.

Exercises

Here are some steps in organic mechanisms. In cases where arrows are supplied, write the products taking particular care about formal charge. In cases where products are given supply the appropriate arrows.

111.

_____

112.

_____

113.

_____

114.

115.

_____

116.

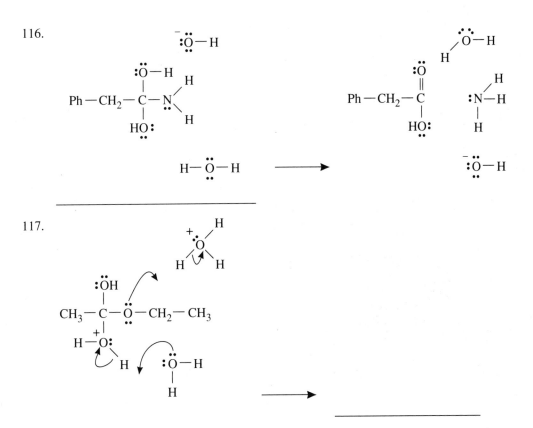

_____

117.

_____

Rearrangements

The group of reactions in which rearrangements occur is one feature that lends organic chemistry its pizzazz. For some, discerning what happened during a rearrangement presents a stimulating challenge; for others, it is a complete mystery. In all cases the rearrangement can be figured out by pushing electrons properly and by carefully keeping track of the relative positions of the atoms.

The first, and most common, rearrangements encountered in the study of organic chemistry are 1,2-shifts in carbocations. In these reactions, a group (Y) one carbon atom away from a carbocation center moves, with its pair of electrons, to that center. The carbon atom from which that group migrated is left with a formal positive charge.

In this general example the pushable electrons are the sigma electrons of the C–Y bond, and the receptor is the positively charged carbon atom. The driving force for these 1,2-shifts is usually the formation of a more stable carbocation. The groups which most often take the role of Y are hydrogen, methyl, and aryl.

The 1-butyl cation suffers a 1,2-hydride shift to become the more stable 2-butyl cation. A neighboring hydrogen atom, with its pair of sigma electrons, migrates to the positively charged carbon atom, which becomes neutral. The neutral carbon atom from which the hydride ion migrated obtains a positive charge.

Notice, as you learned in Chapter 2, that *electrons* are pushed, not the positive charge.

118. The isobutyl cation rearranges via a hydride shift to the *t*-butyl cation.

_____   _____

supply arrow

119. A 1,2-shift of a methyl group is illustrated in the rearrangement of the 3,3-dimethyl-2-butyl cation to the 2,3-dimethyl-2-butyl cation.

_____

120. Similarly,

_____   _____

supply arrow

*   *   *   *   *

In each of the cases above the rearrangement resulted in the formation of a more stable cation, that is, a secondary and tertiary cation are formed from a primary, and another tertiary cation is formed from a secondary. Very often one must choose the most likely rearrangement from among several alternatives.

121. The 1-propyl cation can rearrange via either a 1,2-hydride shift or a 1,2-methyl shift. The 1,2-hydride shift leads to the more stable isopropyl cation.

_____   _____

supply arrow

A 1,2-methyl shift is not as likely because it would yield a primary cation as the product of rearrangement.

122. Show the most likely rearrangement for the 2-methyl-1-butyl cation.

_____   _____

supply arrow

123. In the cation below there are three different groups (hydrogen on C_1, hydrogen on C_3, or methyl on C_3) that can shift to give three possible products of rearrangement. Show the most likely rearrangement.

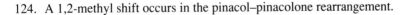

_____   _____

supply arrow

124. A 1,2-methyl shift occurs in the pinacol–pinacolone rearrangement.

_____   _____

supply arrow

125. Groups can migrate from one position on a ring to another.

———————————————   ———————————————

supply arrow

126. When a phenyl group is found next to a cationic center, there is a strong possibility that it will migrate.

————————————————

supply arrow

127. Phenyl groups can migrate from one position on a ring to another.

———————————————   ———————————————

supply arrow

128. Phenyl groups migrate in the pinacol–pinacolone type rearrangements.

———————————————   ———————————————

supply arrow

129. Electron releasing substituents on the aryl ring tend to make that ring migrate in preference to an unsubstituted phenyl group. In the reaction below, the *para*-methoxyphenyl group migrates in preference to phenyl.

supply arrow

130. Electron attracting groups have the opposite effect so that phenyl migrates in preference to *para*-chlorophenyl.

supply arrow

Exercises

Here are some carbocations that can be expected to rearrange via 1,2-shifts. Show the rearrangement which is most probable.

131.

$$CH_3-\underset{\underset{CH_3}{|}}{\overset{\overset{CH_3}{|}}{C}}-\overset{+}{CH}-\underset{\underset{CH_3}{|}}{\overset{\overset{H}{|}}{C}}-H \longrightarrow$$

132.

_____   _____

133.

_____   _____

134.

_____   _____

135.

_____   _____

136.

*   *   *   *   *

Another 1,2-shift involving carbocations is very similar to the rearrangements we have already seen. However, it often tortures students because it involves ring expansion or ring contraction that is difficult to see at first. In the general example below, a carbocation outside the ring is a receptor for a migrating methylene group. The result of such a migration is, inevitably, a ring containing one more carbon than the original ring. The positive charge is now on a carbon atom in the ring.

The cyclohexylmethyl cation rearranges by expanding to the cycloheptyl cation. The author has found it useful to show the rearrangement of the bonds while keeping the positions of the atoms constant. Then, after the rearrangement has been shown, the structure can be altered to a more familiar form.

137. The cyclopropyl methyl cation suffers a similar rearrangement.

138. Similarly, the cyclobutylmethyl cation undergoes a ring expansion.

supply arrow

139. Another example.

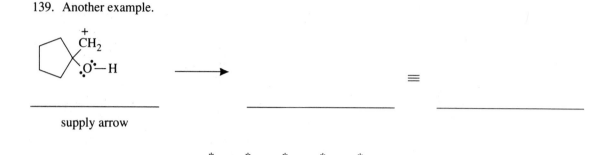

supply arrow

  * * * * *

Alternatively, the carbocation center can be inside the ring. In this case the migration of a methylene group leads to a ring contraction. The ring will contain one fewer member than the original ring, and the positive charge will be on a carbon atom outside the ring.

  The 2,2-dimethylcyclohexyl cation undergoes a ring-contracting rearrangement. Once again it is useful to keep the atoms in the same position as the bonds are rearranged.

140. The cyclopentyl cation will contract to the cyclobutylmethyl cation.

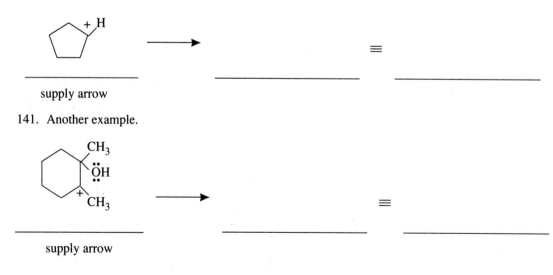

supply arrow

141. Another example.

supply arrow

142. A slight variation on this reaction occurs when cyclopropyl cations rearrange to form allyl cations. Scrutiny of this process will show it to be like a ring contraction, the double bond being comparable to a two-membered ring.

_____

143. Another example.

_____   _____   _____

   supply arrow

Ring Closures

The trick of holding atoms in place while pushing electrons works well in another class of reactions that sometimes vexes students. Consider this remarkable transformation.

$$H_2C=CH-CH_2-CH_2-CH_2-\overset{..}{\underset{..}{O}}-H \xrightarrow[H_2O]{Br_2} \overset{..}{\underset{}{O}} \quad CH_2-\overset{..}{\underset{..}{Br}}:$$

$$\quad\quad\; 5 \quad\; 4 \quad\;\; 3 \quad\quad 2 \quad\quad 1$$

After rejecting the notion that a miracle has occurred, the wise student works on three premises: familiar chemistry is at work, most of the atoms in the starting material will appear in the product, and a minimum number of bonds have been broken and made.

Study how the product relates to the starting materials. There are five carbons and one oxygen in each. Connecting the oxygen with C_4 will give a 5-membered ring. Carbon-4 is part of a double bond. Bromine reacts with double bonds.

$$H_2C=CH-CH_2-CH_2-CH_2-\overset{..}{\underset{..}{O}}-H \longrightarrow \overset{+}{CH_2}-CH-CH_2-CH_2-CH_2-\overset{..}{\underset{..}{O}}-H$$

$$:\overset{..}{\underset{..}{Br}}-\overset{..}{\underset{..}{Br}}: \quad\quad\quad\quad\quad\quad\quad\quad\quad\quad :\overset{..}{\underset{..}{Br}}:$$

Now there is a way for oxygen to bind to C_4. Note that, temporarily, all the atoms are held in place.

Now the structure with the bizarre, elongated bond is converted to the more realistic and familiar form.

Loss of a proton from oxygen gives the product.

Here are some ring-closure reactions. Some are homolytic; others are heterolytic. Where arrows are supplied, draw the products. Where products are given, supply arrows.

144.

_____

145.

$$CH_2=CH-CH_2-CH_2-CH_2-\overset{\cdot}{C}H_2 \longrightarrow$$

_____

146.

_____

147.

$$:\overset{-}{\overset{..}{O}}-CH_2-CH_2-CH_2-CH_2-\overset{..}{\underset{..}{Br}}: \longrightarrow \quad + \quad :\overset{..}{\underset{..}{Br}}:^-$$

_____

148.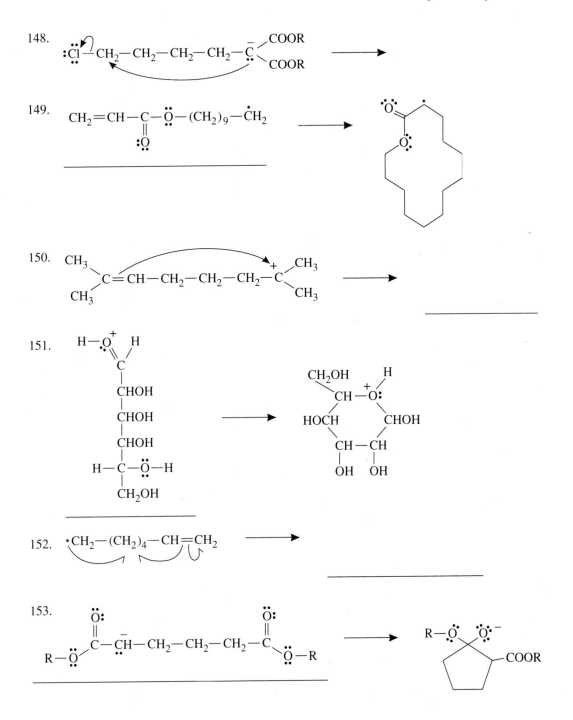

149. $CH_2=CH-C-\overset{..}{\underset{..}{O}}-(CH_2)_9-\overset{.}{C}H_2$ ⟶

     $\underset{:O:}{||}$

     _____

150.

151.

     _____

152. $\cdot CH_2-(CH_2)_4-CH=CH_2$ ⟶

     _____

153.

     _____

Some Reactions From Biochemistry

As a final excercize, I present some steps in reactions that are important in the study of biochemistry. Where arrows are supplied, draw the products. Where the products are given, supply arrows.

154.

155.

156.

157.

158.

Answers

Chapter 3

1. chloride, carbon–chlorine, chlorine, negative, positive, isopropyl,

2. bromide, carbon–bromine, bromine, negative, positive, bromide, cyclohexyl,

3.

4. oxygen, neutral, positive,

5. oxygen positive, neutral, positive,

6.

7.

8.

9.

10.

11.

12.

13.

14.

15. unshared, bromide, carbon, bromine, negative, neutral

16.

17. unshared, nitrogen, carbon, sigma, positive, positive, neutral,

18.

19.

20.

21.

22.

23.   hydroxide, chloride, unshared, sigma, chlorine, carbon, chlorine, neutral, negative,

24.

25.   hydrogen, oxygen,

26.   unshared, nitrogen, sigma, carbon, hydrogen, carbon, neutral, negative,

27.

28. propyl, unshared, carbon, sigma, oxygen, oxygen, carbon, oxygen, neutral, negative,

29. ethoxide, unshared, oxygen, carbon, sigma, oxygen, oxygen, carbon, oxygen, neutral, negative,

30. cyclohexanol, oxygen, hydronium, unshared, hydrogen, sigma, oxygen, oxygen, hydrogen, oxygen, positive, neutral,

31. oxygen, anilinium, unshared, hydrogen, sigma, nitrogen, nitrogen, hydrogen, nitrogen, neutral, positive,

32.

33.

34.

35.

36.

37.

38.

39.

40.

MgBr

41.

42.

43. CH₃—CH₂ •   H—C̈l:

44.

45. H₃C — B̈r: • B̈r:

46.

47.

48.

49.

50.

51.

52.

53.

54.

55. hydroxide, unshared, oxygen, oxygen, negative

56. hydroxide, unshared, pi, oxygen, oxygen, carbon, oxygen, neutral, negative,

57.

58.

59.

60. carbon, oxygen, positive, neutral

61. unshared, oxygen, pi, oxygen, carbon, oxygen, positive, neutral

62.

63. pi, carbon, carbon, neutral, positive,

64.

65. pi, carbon, hydrogen, neutral, positive,

66.

The alternative,

is less likely since it results in a secondary carbocation.

67. carbon, oxygen

68.

69. carbon, oxygen, positive, neutral

70.

71. carbon, carbon, positive, neutral

72. carbon, carbon, carbon, positive, neutral

73.

74. carbon, carbon, positive, neutral

75.

76.

77.

78.

79.

80.

81.

82.

83.

84.

85.

86.

87.

88.

89.

90.

91.

92.

93.   (a)

(b)

94. (a)

(b)

95. $Cl_3C - CH_2 - \overset{\bullet}{C}H - CH_2 - CH_2 - CH_3$

96.

97.

98. (a)

(b)

99.

100.   hydrogen, carbon, neutral, negative,

101.

102.   neutral, neutral,

103.

104.

105. , positive, negative

106.

107.

108.

109.

110.

111.

112.

113.

$:O:$
$\parallel$
$Ph-C-\overset{\cdot\cdot}{N}-Ph$
$\vert$
H

$:\overset{\cdot\cdot}{O}-H$
$\vert$
H

114.

115.

116.

117.

118.

119.

CH₃—C—CH—CH₃

120.

121.

122.

123.

124.

125.

126.

127.

128.

129.

130.

131.

132.

133.

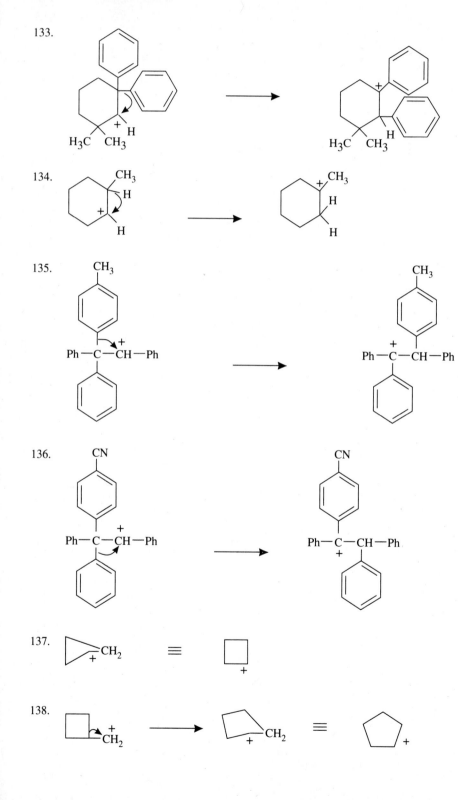

134.

135.

136.

137.

138.

139.

140.

141.

142.

143.

144.

145.

146.

147.

$$:\overset{..}{\underset{..}{O}}{}^{-}-CH_2-CH_2-CH_2-CH_2\overset{\frown}{\,}\overset{..}{\underset{..}{Br}}:$$

148.

$$\text{(cyclopentane ring)}\begin{matrix} COOR \\ COOR \end{matrix} \quad + \quad :\overset{..}{\underset{..}{Cl}}:^{-}$$

149.

$$CH_2\!=\!CH-\overset{O}{\underset{\|}{C}}-\overset{..}{\underset{..}{O}}-(CH_2)_9-\overset{.}{C}H_2$$

150.

$$\begin{matrix} CH_3\,CH_3 & CH_3 \\ \text{(cyclopentane ring)}\!-\!\overset{+}{C} \\ CH_3 \end{matrix}$$

151.

$$H-\overset{+}{\underset{..}{O}}\!=\!\overset{H}{\underset{\substack{C \\ | \\ CHOH \\ | \\ CHOH \\ | \\ CHOH \\ | \\ H-\overset{..}{C}-\overset{..}{\underset{..}{O}}-H \\ | \\ CH_2OH}}{}}$$

152.

(cyclohexane ring)$-\overset{.}{C}H_2$

153.

$$R-\overset{..}{\underset{..}{O}}\overset{\overset{..}{\underset{..}{O}}:}{\underset{\|}{C}}-\overset{-}{C}H-CH_2-CH_2-CH_2-\overset{\overset{..}{\underset{..}{O}}:}{\underset{\|}{C}}\overset{..}{\underset{..}{O}}-R$$

154.

155.

156.

157.

158.